Gravity of the Matter

Understanding Gravity

By Mark F. Dennis

A Golden Iris LLC Publication

2014

Illustrations by
Mark Dennis

Cover Image of the Cat's Eye Nebula courtesy of
NASA and
http://www.spacetelescope.org/images/opo9501a/

The Cat's Eye Nebula - NGC 6543

iv

"To deny the aether is ultimately to assume that empty space has no physical qualities whatever. The fundamental facts of mechanics do not harmonize with this view...

...besides observable objects, another thing, which is not perceptible, must be looked upon as real, to enable acceleration or rotation to be looked upon as something real...

...the conception of the aether has again acquired an intelligible content, although this content differs widely from that of the aether of the mechanical wave theory of light."

Albert Einstein, Lecture in Leiden (1920)

(Isaacson, pg. 318)

Table of Contents

Table of Figures

Gravity of the Matter

Understanding Gravity

This book has been condensed from my 2010 book "*A Challenge to Einstein*" which covered much of Einstein's theories such as Special Theory of Relativity and the General Theory of Relativity. This book focuses primarily on the theory behind gravity and its properties.

Gravity of the Matter

Introduction

The properties of gravity have always amazed humans, even from before the days of Plato, Aristotle and Socrates. What causes this attraction at a distance and how can light waves and the forces such as gravity and magnetism operate across the vast emptiness of space? If the pull of gravity reaches out millions of miles through empty space, what is it that is doing the pulling? Could it be that mass exists simply as energy manifested as standing waves within some sort of rigid elastic substance? How was mass created? What is gravities relationship to mass? What relationship does time have with gravity? Is the universe really accelerating outward or is it collapsing inward? These were questions that drove the genius of Albert Einstein to doubt the scientific theories that were in place in the early twentieth century and develop new ideas of his own. Now, almost 100 years later, Einstein's theories have become the foundation of quantum mechanics and modern theoretical physics. Today in the early twenty-first century, is it not time to question the old theories again?

These are all questions that we shall discuss in this book and hopefully, if anything, understand a little more about the universe and the world in which we live, the big and the small. A great deal of effort has been taken to explain the mind and thoughts of Einstein and many other brilliant thinkers throughout history in simple everyday language that the layman and even the young student can understand. Reading this book will open your scientific mind!

Gravity of the Matter

Propagation of Energy

Before the year 1900, most scientists believed that energy was propagated through space in what was called the "Aether". They believed that like as sound moves through air, the energies of light, magnetism and gravity must travel through something, even across the vast emptiness of space. The belief was that the energy-bearing "aether" was a fine gas-like substance made of the smallest of particles.

Ancient Greeks and much of the world at that time believed that all things were made of combinations of the four basic substances, fire, water, earth, and air. Among the ancient Greeks were Greek philosophers Plato (427-347 BC) and Aristotle (384 - 322 BC), born in the height of ancient Greek civilization. Plato, a student of Socrates (469–399 BC), was a philosopher and mathematician who laid the foundations of modern philosophy. Aristotle is believed to be the first person to attempt physics and even gave physics its name which means the "study of matter" and its motion. Aristotle is one of the central figures of the era and was a student of Plato (Bord, pg. 33).

Aristotle and Plato before him believed that in addition to the four basic elements there was a fifth element which Aristotle would call the aether, meaning "to blaze" (Ross, pg. 461). This aether would be the main constituent of the celestial bodies (Asimov, pg. 6). It would be a substance rigid enough to transmit energy at incredibly high speeds while offering no resistance to the movement of the galaxies, stars and planets throughout the universe

1

(Barbour, pg. 128). This was the belief for thousands of years until around the turn of the 20th century when new scientific discoveries changed things. Throughout this discussion we will at times call what is sometimes referred to as the aether, a "*field*" or a "*medium*".

Galileo (1564-1642) lived in Italy during the Renaissance, one of the most productive eras in human history. He was born the same year as Shakespeare and born the same year that Michelangelo died. Galileo gave science the gift of many discoveries in both mechanics and astronomy, such as the sun centered heliocentric model of or solar system, which led to him being placed on trial by the Roman Inquisition. Galileo also discovered the law of falling bodies. Aristotle, hundreds of years earlier had reasoned, erroneously, that the velocity of falling objects were dependent on their weights. Galileo would later discover that the velocity of two objects of different weights fell at the same rate. Legend has it that this discovery was made from the Leaning Tower of Pisa (Bord, pg. 35).

Ole Rømer (1644-1710) was born two years after Galileo's death and was one of the first scientists to discover the speed of light through studies of the movement of one of Jupiter's moons. Romer was a student of Italian mathematician and astronomer, Giovanni Cassini, who discovered of the Great Red Spot on Jupiter in 1665.

Jupiter takes longer to orbit the sun than the earth does (twelve earth years), so that at different times of the year, the earth in its orbit around the sun moves at a high velocity toward Jupiter. At other times of the year, the earth moves away from Jupiter at a high rate of speed. Rømer would build on Cassini's observations during his studies of Jupiter, that the times between Jupiter's moon Io's eclipses got shorter as Earth approached Jupiter, and longer as Earth moved farther away. In 1676, after studying about 140 eclipses of Io, Rømer calculated that the speed of light must be 299,792 km/sec or 186,290 miles/sec (Born, Max pg. 91). We will examine this later.

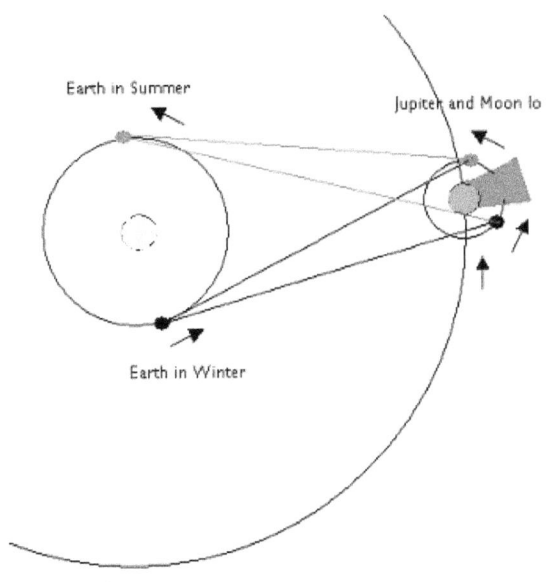

Jupiters Moon Io

Gravity of the Matter

Star Aberration

During the eighteenth century, the Copernican heliocentric view of the universe had generally been accepted although there was very little empirical data to support it (Born pg. 10). It was for this that Galileo in 1633 had risked martyrdom for the sake of what he believed to be truth.

In December 1725, one year before Newton's death, English astronomer James Bradley (1693-1762) began a prolonged observation of Gamma Draconis (also known as Eltanin), a star in the Draco constellation. This constellation is a group of circumpolar stars that remain in the night sky all year around and which passes directly overhead nightly where he lived in London. The study was an effort to the view the parallax of this star, using triangulation to determine its distance from the earth.

Throughout his observation viewing Gamma Draconis through a telescope fixed to the chimney on the roof, Bradley found that he had to continuously adjust his telescope to a different position almost nightly. Over a six months period, the position of the star seemed to move one direction then the other from its central position with a maximum variation of approximately 20.5 arc seconds. Unlike the circular motion expected due to the earth's tilt on its axis, the star seemed to move in an elliptical pattern. Bradley discovered that this angle of "aberration" depended on the ratio between the orbital velocity of the earth and the speed of light coming from the star.

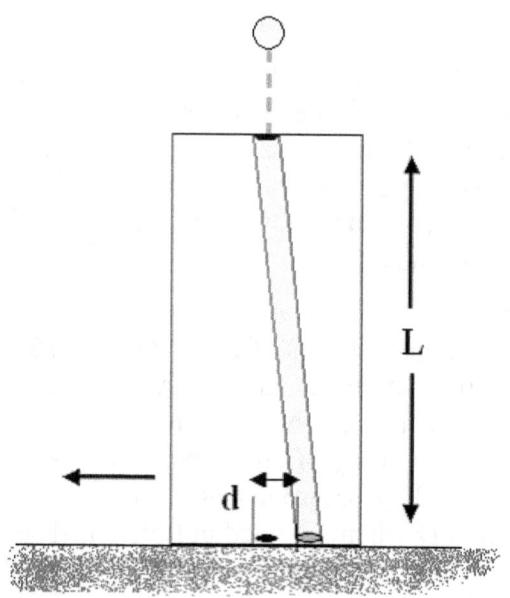

Star Aberration

Another way to understand this is through the movement of the earth in orbit around the sun. It takes light 8.3 minutes to travel from the sun to the earth, therefore, where you see the sun in the sky is where the sun was about 8.3 minutes ago; in reality the sun's true position is about 20 arc seconds from its visible position. However, when measuring the acceleration vector for the Earth's motion in orbit around the sun and comparing the gravitational position of the sun to the apparent visual position due to transit delay, there is a difference. If gravity propagated through space at the speed of light, there would have been no difference.

6

Gravity therefore, unlike the electromagnetic forces of light, radio waves, electromagnetic fields and magnetism, has no detectable aberration or delay. If gravity traveled at the speed of light, the earth would tend to rotate around the sun's apparent position and not its true position causing the earth's orbit to change slightly in distance on each rotation. Gravity therefore must propagate at much higher velocities than light, possibly even instantaneously in time and cannot be considered as a force of nature that propagates. Simply stated, gravity must purely be a geometric effect of curved space and time (Space-Time). The thought then comes to mind, if space is empty, what is being curved?

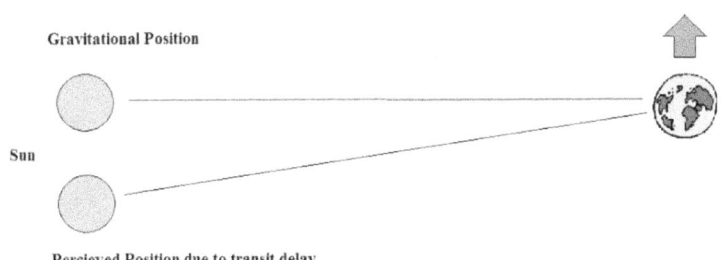

Geometric Effect

Static charges that attract and repel act in much the same way and do not propagate or take time to affect other objects. This seems to indicate that an invisible field surrounds and moves with all particles of mass and instantaneously causes what Einstein called a "spooky" action at a distance (Isaacson, pg. 458).

7

Gravity of the Matter

Like Charges Repel

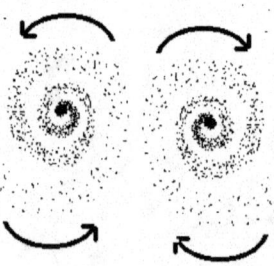

Opposite Charges Attract

Star Dust

In 1644, French philosopher René Descartes (1596-1650) revived the idea of old that the aether was the carrier of energy through space as well as developed the law of refraction, also known as Descartes' law. Descartes' aether would be an elastic medium made up of the smallest of particles, the "God Particle" or "stardust" if you will, permeating everything; space was not empty. This would be a substance that oscillates, setting in motion other particles in waves, moving its energy through "empty" space. Diffraction, refraction and reflection of light all seemed to support this hypothesis that light moved in waves through this aether; why not gravity and magnetism as well? This aether however could not impede the motions or cause resistance to the planets or stars as they moved through it and yet must be rigid enough to transmit light (Barbour, pg. 128). Although the aether would not induce drag on these bodies, it would still be the medium in which the force of gravity would work through.

Descartes undulation or wave theory, hypothesized that this aether would have to be considered an elastic solid due to the fact that light is transmitted as a transverse wave. Transverse waves cannot propagate through the fluidic motions of liquids and gases such as ocean waves and sound wave through air; this is because there is no resistance to lateral displacements of particles in these waves. Solid bodies propagate waves longitudinally and transversely; fore and aft, left and right and up and down, even diagonally as evident with the polarization of light (Isaacson, pg. 115).

9

According to the wave theory, light would theoretically travel through the aether in waves, very much like sound through a steel girder. A wave is defined as:

> *...a traveling disturbance of coordinated vibrations that transmit energy through a substance without net movement of that substance.*

There are two distinct types of waves; longitudinal waves and transverse waves. Both waves can travel in elastic solid mediums, however, waves traveling through fluids such as liquids and gases are always longitudinal waves. Longitudinal waves are waves in which oscillations are along the direction of wave travel such as in sound waves where there is a back and forth motion of these particles. The absence of rigid bonds between particles in a fluid prevents traverse waves from traveling through them. A transverse wave is a wave in which oscillations are perpendicular to the direction of wave travel such as a wave on a rope tied to a doorknob, electromagnetic waves and seismic waves traveling through the ground needing a somewhat rigid medium in order to conduct their energy (Tipler, pg. 465).

Transverse Wave

One way to view a longitudinal wave is to string single particles along in a line, somehow tied together with spring like forces. As one particle is forced away from another, an effect is felt on the particles to its left and to its right. One is pulled along for the short distance it moves and the other is pushed. A chain event takes place as one particle pulls on another and one pushes on the other down the line in one direction longitudinally, also known as a compression wave.

Longitudinal Wave

Now, visualize if you will, not only fore and aft but up and down and left and right motions in which all particles effect each other in all directions

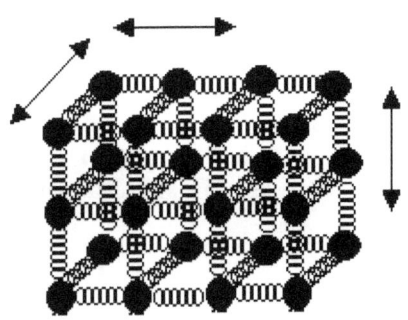

Transverse Wave

Polarization is the orientation of these transverse waves in light waves. Transverse light waves not only are propagated in a certain direction at the speed of light, they also oscillate perpendicularly to the direction of propagation, vertically and horizontally at the speed of light (Bord, pg. 326). The fact that light waves show properties of polarization reveals that light waves are indeed transverse waves. Therefore if light is propagated through a medium, and transverse waves can only be transmitted through solids, then this medium that light moves through must be an elastic solid like an "elastic jelly" (Psillo, pg. 131).

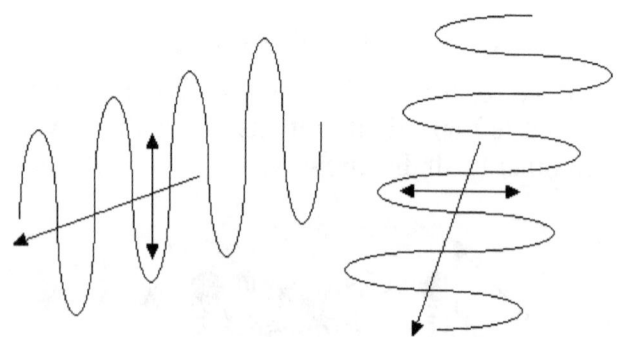

Horizontally and Vertically Polarized Transverse Waves

In September of 1818 French physicist Augustin Fresnel, (1788-1827) sent this letter to François Arago:

"By your fine experiments on the light from stars, you have demonstrated that the movement of the terrestrial globe has no noticeable influence on the refraction of rays emanating from these stars..."

"You have inspired me to investigate whether the result of these observations could be more easily reconciled with the system which takes light as consisting of vibrations in a universal fluid. It is all the more necessary to explain these results using this theory, since it must apply equally to terrestrial objects; for the speed with which the waves are propagated is independent of the movement of the bodies from which they emanate."

(Fresnel, pg. 57–66)

In 1887 - Albert Michelson (1852-1931) and Edward Morley (1838-1923) set out to prove the existence of the aether using an interferometer. It was assumed by the experts of the day that the aether was a rigid substance, which was stationary in space and was the *medium* in which light traveled. It was believed that as the stars and planets traveled through space, they passed through this substance creating an "aether wind" that should be detectable with some sort of device. Michelson and Morley set up an interferometer (a devise that utilized mirrors and detectors to measure the speed of light) in such a way that as the earth passed through the aether wind in space,

13

distortion or interference fringes would develop on the detector. The theory was that the speed of light would be different in each of the arms of the detector, one pointing into the aether wind (the direction the earth was moving in space) and the other perpendicular to the assumed aether wind. No wind was detected. (Ditchburn, pg. 315).

Michelson's Interferometer

The famed Michelson Morley Experiment in 1887 with the *interferometer*, and other experiments involving light wave interference phenomena confirmed to scientists of that day that there was no aether. Not only did they **seem to show that there was no aether**, they had also discovered what they believed to be the "Law of the Constancy of the Speed of Light." This constant is expressed as "c" and is measured at 299,853 km/sec. or 186,328 miles/sec. This measurement of the speed of light in air would not vary and is the same velocity no matter what the observer's velocity is or what velocity of the emitter is, either moving towards or away from each other (Wolf, Fred, pg. 52).

Gravity of the Matter

Partial Aether Drag Theory

"If one supposed that our globe impresses its motion on the aether in which it is enveloped, one would easily conceive why the same prism always refracts light in the same manner, whatever the side from which the light may arrive. **But it seems impossible to explain the aberration of stars by this hypothesis***: I have not been able, so far at least, to account for this phenomenon clearly except by supposing that the ether passes freely through the globe, and that the speed communicated to this subtle fluid is only a small fraction of that of the earth, not exceeding one hundredth of it, for example"*

(Fresnel, A. pg. 57–66)

The Michelson Morley Experiment seemed to indicate that there was no such thing as the aether. Fresnel however suggested that the aether or medium through which light and gravity traveled, must have been sticking to the earth, being dragged along. As light from the stars would enter this medium at different velocities, it would enter the earth's atmosphere at one speed. It seems however, that if the aether drag hypothesis were true, then stellar aberration could not occur because the light would be traveling in the aether, which would be moving along with the telescope (Van der Kamp, Walter pg. 30). Fresnel at first saw this as contradictory but then suggested that material with low refractive indexes, such as air, must produce almost no

17

resistance at all to stellar light allowing stellar aberration to occur. However, substances with a higher refractive index, such as water and glass, produce more resistance, thus allowing for Arago's experiment that indicated the constancy of the speed of light.

Light would therefore travel unimpeded through the air and be uninfluenced by the movement of the earth. Once light enters into glass or some other medium of higher refractive indexes, it would therefore be constrained or locked into the new medium and travel at a different (slower) velocity within that medium. The Fresnel **partial drag theory** was widely received during the last part of the nineteenth century. This both allowed for the constancy of the speed of light and for star aberration.

The partial aether drag theory suggested that the velocity of light would be affected by a moving medium. Fresnel projected that the aether's density in a substance was different than that of the aether in empty space and depended on the refractive index (n) of the substance it was traveling through. Air would have very little resistance to the aether but substances like water and glass would have a lot of resistance. Here c = the speed of light, n = refractive index, and v = to the velocity of the medium (Ditchburn, pg. 334).

$$c' = c_1 + v\left(1 - \frac{1}{n^2}\right)$$

Fresnel's Convection (Aether Drag) Coefficient

Partial Aether Drag Theory

In an effort to find some variation in the stellar aberration, Sir George Biddell Airy in 1871 carried this idea a little further with the use of a water-filled telescope as suggested by Fresnel in his 1818 letter to François Arago. Theoretically as light would travel through the water as compared with air, refraction of light in the water would affect stellar aberration in some degree but it did not.

Gravity of the Matter

The Dilemma

The speed of light was now known. However, experiments such as the famous Michelson – Morley Experiment indicated that the velocity of light was always the same and independent of the motion of the emitter or the observer. The "simple emission theory" suggests that light given off by an object should move at a velocity relative to the movement of the emitting object and/or the movement of the observer, however this was not the case. The aether drag theory tried to answer this with an aether that was being dragged along with the earth acting as a medium, causing all light from heavenly bodies to be perceived at one velocity – like sound is through air. Stellar aberration however contradicted this by proving that light was not being carried by any medium, at least not in the earth's atmosphere. Fresnel partial drag theory tried to correct this by stating that aether drag was dependent on refractive indexes and that air had little effect on light; denser object such as glass and water however would carry light at varying degrees. Airy's water filled telescope experiment in 1871 seemed to contradict this.

In 1913 while studying Binary Stars or twin stars that orbit around each other, Willem de Sitter (1872-1934) a Dutch mathematician, physicist and astronomer, suggested that stars in these systems should have an orbit that emitted light at different speeds (c + v or c − v). The problem was that the "faster" light given off during the orbit of a star towards the earth would be able to catch up with and even overtake the "slower" light emitted from the other star in the system retreating away from the earth. Theoretically, this difference in light velocities would cause the stars light to be scrambled and out of sequence, distorting the light seen from these binary stars (Einstein, Albert, pg. 18).

The fact that there are no distortion means that we can be positively sure that the speed of light coming from both stars is the same from the instant it left the stars. With the assumption that the movement of observer is equivalent to the movement of the stars, this leads us to the conclusion that the speed of light does not depend on the speed of observer or the movement of the stars; the speed of light will always be 300,000 km/sec to the observer detecting the light (Sokolov, G. pg. 2).

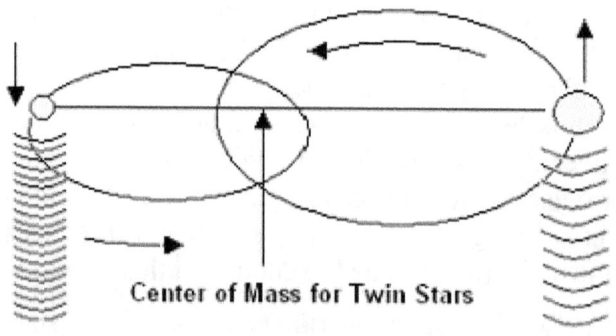

Center of Mass for Twin Stars

Doppler Effect Twin Stars

Although the light coming from these stars travels at the same velocity, the wavelength of that light is not the same. When one of the twin stars is moving towards the earth, the wavelengths compress and become shorter, towards the blue in the light spectrum, known as Blue Shift (similar to the effect of a whistle emitted from the caboose to an observer on the train). When one of the twin stars is moving away from the earth, the wavelengths are shifted to longer wavelengths into the red portion of the light spectrum, known as Red Shift (similar to the effect of a whistle emitted from the engine to an observer on a train) producing the "Doppler Effect" (Bless, pg. 235).

This now creates a dilemma and raises some important questions. If light is made up of particles called photons or "hot ejected particles" as Newton would put it, and was leaving a star <u>moving towards</u> a detector or an observer at the speed of light, what is it that slows the photons down? If the star emitting the light being detected is <u>moving away</u> from the detector, what speeds up these photons? The photons must be moving at the speed of light in reference to the moving star that they came from yet be moving at the speed of light in respect to the detector or observer. How can this be?

Before the 1900's it was natural for almost everyone to believe that light must be transmitted through the material called the aether, which permeated everything. This aether would fill all of the space between the stars and the earth; otherwise light could not be visible coming from these stars. If light moved in waves as it seemed to, especially transverse waves, just as there must be air for sound to move through, there must also be a medium for light to move through. This would be true for not only light but gravity, radio waves and even static charges as well.

In his book **Opticks**, published in 1704, Sir Isaac Newton wrote:

> "*Is not the heat of the warm room conveyed through the vacuum by the vibrations of a much subtler medium than air, which after the air was drawn out remained in the vacuum? And is not this medium the same with that medium by which light is refracted and reflected, and by whose vibrations light communicates heat to bodies, and is put into fits of easy reflection and easy transmission?*"

(Newton, pg.323)

Now we do know that sound has a fixed velocity in air at a given temperature; would this also be the case with light traveling through this aether? If this aether was present throughout space, as the Earth rotates on its axis and moves in its orbit around the sun, should not this aether be possibly detected as an eerie **wind** blowing through the earth, affecting the wavelengths of light moving in specific directions. Light waves moving with the earth's rotation it seems should be compressed and light waves moving perpendicular to the earth's rotation should not be affected.

As previously discussed, the theory behind the interferometer built by Michelson and Morley was built on this same theory. As the earth rotated and moved in its orbit around the sun, it was believed that this theoretical aether wind that was stationary in space would flow through the detector at a specific angle. A light beam

24

would be split with a half-silvered glass plate and be reflected off two mirrors. One beam would be pointed into the aether wind (presumed to be flowing in the opposite direction of the earth's rotation and movement through space) and the other reflected perpendicular to the aether wind. The light beamed into the aether wind would slow down compressing the waves, making them shorter. When these waves enter the detector they would be out of phase with the perpendicular beam that was not affected causing distortion on the detector. If there was no distortion there evidently was no aether wind.

The detector was built and floated on pool of liquid mercury to dampen out any noise and vibrations caused by the surroundings. After several experiments with no distortion in the detector, it was determined that there was **no aether wind** (Collins, H. pg. 37). The Michelson Morley experiment seemed to confirm that light takes the same time to travel each path irrespective of the motion of the observer or detector.

The Michelson Morley Experiment

Gravity of the Matter

In 1895, Dutch physicist Hedrick Lorentz (1853–1928) theorized that the "null" result obtained by Michelson and Morley must have been caused by an "effect of contraction" made by the aether on their apparatus and introduced the length contraction equation (Collins, H. pg. 38). Where L is relativistic length and L_0 is length of object at rest. The idea here is that the aether must still exist but that high-speed movement through space distorts space enough to nullify any measurements that could detect the aether; the higher the velocity the more contraction of space. This is now known as the "Lorentz Contraction". These discoveries at the turn of the century changed science; from this, Albert Einstein developed his "Theory of Relativity".

$$l' = l \sqrt{1 - \frac{v^2}{c^2}}$$

Lorentz Contraction Equation

Albert Einstein (1879-1955) was born in 1879 into a nonreligious Jewish family in the city of Ulm, southern Germany (Isaacson, Walter pg. 11). He was born the same year of Scotsman James Clerk Maxwell's death, one of the leading scientists of the day in the field of Electricity and Magnetism, an area in which Einstein would become well versed. Apparently stubborn, young Einstein did not speak until the age of 3, waiting till he could talk in full sentences. He loved playing with puzzles, working geometry problems and enjoyed toying with magnets. Later on as a teenager, due to conflict with his teacher,

26

The Dilemma

Einstein would end up dropping out of high school, but was still able to enroll in a Swiss university. Although being considered "lazy" by some of the professors there, Einstein soon became well-schooled in the area of physics; a field he would soon pioneer and later would develop the famous "Theory of Relativity" (Bartusiak, pg. 29-30).

Gravity of the Matter

The Special Theory of Relativity

Albert Einstein began to wrestle with these seemingly contradictory findings, struggling with the properties of light whether it was wave or particle and the constancy of the speed of light. After years of study, he came to a conclusion that if the velocity of light was constant, no matter what the velocity of the observer or the velocity of the object emitting the light (as indicated by the famous Michelson – Morley Experiment in 1887) that the properties of both time and space must be changing.

In order to fully understand Einstein's "Theory of Relativity", one needs to understand "relative velocity". To measure the velocity of any object, you must have a means of measuring distance and a means of measuring time. Einstein visualized this in his writings with a measuring rod (yard stick, tape, whatever) and a clock (Kafatos, Menas, pg. 24). Today, in the modern world we can use a baseball and a radar speed detector as an example.

If the baseball is thrown towards the stationary radar detector at 100 mph, the detector will detected the baseball at that same velocity - 100 mph. Now, let's put the radar detector in a vehicle and drive at 50 mph towards the baseball being thrown at 100 mph. This moving baseball can now be viewed as having two speeds, one speed relative to the pitcher who threw the ball at 100 mph and the speed of the baseball relative to the vehicle being detected at 150 mph; the combination of the vehicles speed of 50 mph added to the speed of the ball relative to the pitcher of 100 mph.

Gravity of the Matter

Baseball at 150 mph

On the other hand, if the detector on the vehicle is moving away from the baseball at 50 mph and the baseball itself is thrown at 100 mph relative to the pitcher, the baseball will be detected by the radar detector at 50 mph (as viewed from the moving vehicle) (Bord, 2000, pg. 448).

The velocity of the baseball has to be measured relative to something, in this case, a radar speed detector moving in a vehicle or the velocity of the baseball relative to the pitcher. During the pitch, the ball is traveling at a different velocity relative to the pitcher who threw the ball and relative to the moving radar detector in the vehicle; it's all a matter of perspective. A bystander or passerby will have another relative velocity depending on their direction of movement and velocity relative to the baseball.

Baseball at 50 mph

30

The Special Theory of Relativity

The problem now is that when measuring the velocity of light, no matter what the velocity of the observer (radar gun in this case) towards or away from the light source (baseball), the measurement of the velocity of the light from this object remains the same (Born, pg.231-232).

We can use the earth as an example of this in much the same way as the baseball. As the earth orbits the sun, its mean orbital velocity through space is approximately 30 kilometers/second (18 miles/sec) or 64,800 miles per hour. As the earth in its orbit moves directly toward a star, (assuming that the star is stationary in space) the velocity of the light detected from that star, _should_ be the sum of the velocity of the earth toward the star and the speed of the light from that star, similar to the baseball and radar detector moving towards each other (simple emission theory). On the other hand, as the earth moves away from the star in its orbit around the sun, the speed of the light detected _should_ be the difference between the velocity of the earth in its orbit away from the star and the speed of the light from the star, similar to the baseball and radar detector moving away from each other (Born, pg. 132).

Movement of the earth towards a star should be:

EXAMPLE 1: 186,000 m/sec +18 m/sec = 186,018 m/sec

Movement of the earth away from a star should be:

EXAMPLE 2: 186,000 m/sec - 18 m/sec = 185,982 m/sec

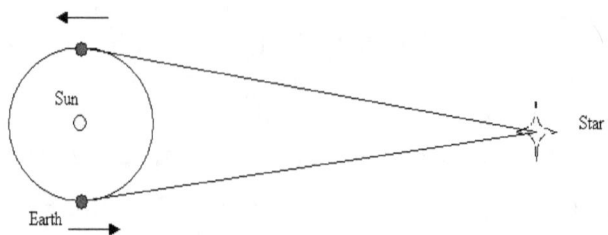

Earth and Star

But this is not the case. Sir Isaac Newton's Classical Mechanics and this new Theory of Relativity are now in conflict.

Einstein himself said:

"Who would imagine that this simple law (the constancy of the velocity of light in a vacuum) has plunged the conscientiously thoughtful physicist into the greatest of intellectual difficulties."

(Einstein, Albert, pg. 19).

"Law of the Constancy of the Speed of Light" was now stated:

"The velocity of the propagation of light cannot depend on the velocity of motion of the body emitting the light" (nor the motion of the observer detecting the light).

(Einstein, Albert, pg. 18).

The Special Theory of Relativity

No matter what the velocity of a detector towards or away from the star, or the velocity of the star itself towards or away from the earth (or detector), the speed of light coming from the star into the detector is always the same, 186,000 m/sec. The math doesn't seem to work – according to Einstein, space and time must be being warped. The question being asked again is, what is being warped?

EXAMPLE 1: 186,000 m/sec +18 m/sec = 186,000 m/second.

Earth Traveling Towards a Star

EXAMPLE 2: 186,000 m/sec - 18 m/sec = 186,000 m/sec

Earth Traveling Away From a Star

To add to this perspective, in addition to the earth's rotation around the sun being 30 km/sec (18 mi/sec) in reference to a star in our galaxy, the movement of the solar system rotating around the center of the Milky Way galaxy is about 250 km/sec (140 mi/s). The Milky Way itself is traveling approximately 300 km/s (185 mi/s.) in reference to other galaxies in the "Local Group". Most galaxies are spreading out away from each other at enormous velocities in an ever-expanding universe. The great Andromeda Spiral Galaxy however, one of the Milky Way's nearest neighbors, is actually moving toward the Milky Way and earth at about 50 km/s (about 30 mi/s.)

The General Theory of Relativity

Albert Einstein would then build on "Special Relativity" with the knowledge he gained in studying gravitational fields. He would go on to explain the universe and everything within it as four-dimensional "Time-Space". In three-dimensional space, the x axis would be the fore and aft, the y axis would be the side to side and the z would be the up and down with respect to the "Fixed System" reference; the fourth dimension is "time", creating Einstein's universe of Space-Time. Velocity could then be explained as the change in position over time. From this perspective, Einstein would develop the theory of the constancy of the speed of light, the theory of gravity as well as explain gravitational redshift and even time dilation.

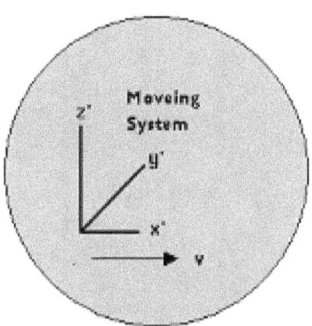

Time- Space

Gravity of the Matter

Gravity as Einstein would see it, exists due to "time-space curvature (Isaacson, pg. 193); as if the fabric of space was being warped. Imagine if you will four people holding a sheet at the four corners tightly, then someone sets down a bowling ball in the center of the sheet causing the center of the sheet to sink. A marble is then rolled onto the blanket; as it rolls, it spins around the bowling ball a few times in a tight spiral until it stops, down between the bowling ball and the blanket. From this premise, Einstein would develop his famous "Field Equation of Gravity":

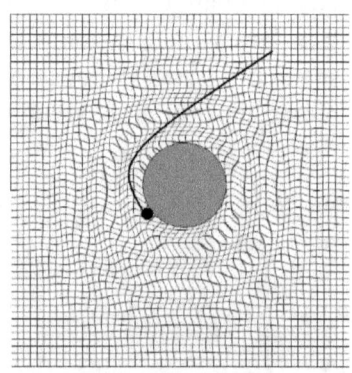

Gravity

Einstein's "Field Equation of Gravity":

$$R_{uv} - \tfrac{1}{2} g_{uv} R = T_{uv}$$

The left side of the field equation of gravity represents a geometric gravity field of space-time and the right, the distribution of mass-energy. Both sides create a balance, the more the mass-energy, the greater the warping of space-time; one affects the other and there cannot be one without the other. Objects such as planets simply naturally follow these curves in time-space, as well does light. Einstein's new formula actually doubled his original formula in the amount of deflection that gravity would cause a ray of light to bend. Later we will see that this was the case; in 1919 when astronomer Arthur Eddington photographed a stars new apparent position during a solar eclipse, where the suns gravity actually changed the stars apparent position in the sky (Bartusiak, pg.45).

We will also see that not only is motion influenced by this curvature of space, but that time also is affected (Einstein, pg. 81). Recent technologies such as the Global Positioning System (GPS) must adjust for time dilation due to the earth's gravitational field and the satellites velocity (Bartusiak, pg. 55).

Gravity of the Matter

Light, Gravity and Relativity

The theory of light prior to Einstein assumed that light was not made of matter and that gravity would have no effect on light. Einstein's Theory of Relativity and the field equations he created generated a new idea, that light was a particle created of matter, gravity therefore attracts light, curving its path as it passes close to a large stellar mass such as a star. As starlight passes close to the sun, the stars apparent position should therefore bend by about 1.7 arc seconds due to the suns gravity (Bartusiak, pg. 47).

Einstein set out to observe and record how much a stars light was deflected by the sun due to the changes in space-time as light passed through its strong gravitational field. This was not an easy task as stars are not normally visible near the sun, so in 1919 astronomer **Arthur Eddington** took on this challenge and traveled to the Isle of Principle off the coast of Africa to test this idea during an eclipse. After studying the photographic plates, it became clear that the stars did shift their apparent position in the gravitational field of the sun. Gravitational forces did bend the rays of the starlight by about 1.6 arc seconds, proving that light was a particle that was made up of mass and that Einstein's theories were holding true (Isaacson, pg. 257).

The following figure shows how the suns gravitational field pushes the apparent position of a star outward from its true position. We also see a similar effect due to gravitational lensing which also changes the apparent position of stars and galaxies due to massive objects throughout the universe.

Apparent Position of a Star

Einstein developed his Principle of General Relativity, a unification of Newton's Law of Universal Gravity and his own Special Theory of Relativity. This theory would go on to predict gravitational lensing, gravitational redshift, the precession of Mercury and many more scientific phenomenon discovered later in the 20th century. This principle was generalized in a mind experiment where passengers on a train moving along with uniform motion feel a sudden change in motion with the abrupt application of the brakes (Einstein, Albert, pg. 73). Einstein would go on to develop the **Equivalence Principle** from this showing that inertial mass that is produced by acceleration/ deceleration and gravitational mass in a gravitational field is essentially the same thing.

There still was a nagging problem; experiment after experiment seemed to prove that the aether was nonexistent and that light consisted of photons that were particles acting as waves. If this aether did not exist, what was waving in reference to light waves? And what did the gravitational, magnetic and electrical fields consist of? A field of what?

Gravitational Lensing

While writing for a science yearbook in 1907, Einstein noted a thought experiment of an observer in a closed windowless chamber that is accelerated upward. He compared the acceleration of this observer to another observer in a gravitational field showing that both felt pressed to the floor and that neither could distinguish the difference, thus formulating his "Equivalence Principle." This principle asserts that inertial mass and gravitational mass are equivalent, that both resistance to acceleration and gravitational effects are manifestations of inertio-gravitational fields.

In this same sort of scenario, an observer is in a chamber accelerating upward, again utilizing the "Equivalence Principle." A pinhole in one wall allows light from outside the chamber to shine on the opposite wall of the chamber. By the time light hits the opposite wall the chamber has moved causing the light to hit the wall a little closer to the floor, in effect bending the light.

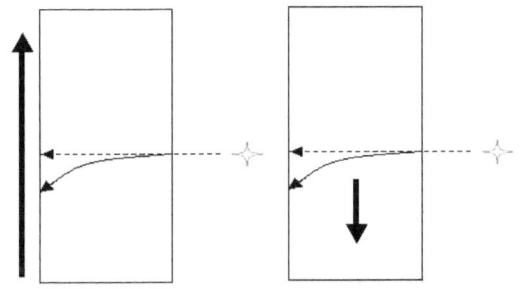

Equivalence Principle

Since inertial effects are equivalent to gravitational effects, gravity too should cause the light to bend (Isaacson, Walter pg. 190). This is strangely similar to James Bradley's Star Aberration reviewed earlier with the "Tall Box Analogy".

Quasars, short for quasi-stellar radio source, were first discovered in the 1950's well after Einstein's theories were developed. Quasars are very powerful electromagnetic wave generators at the center of distant galaxies that are much brighter than normal galaxies. Quasars are the strongest sources of electromagnetic radiation in the universe, possibly generating energy from a super massive black hole at the center of its galaxy. Because of their great distances and brightness, quasars are ideal for studying the effects of gravitational fields on light. Einstein theorized that light would follow a curved path around massive objects due to the immense gravitational field. Today many stellar objects such as Einstein's Cross or Einstein's Rings occur due to the deformation of the light from a source such as a quasar through gravitational or cosmic lensing.

Gravitational lensing occurs when the light from a quasar or other very distant and bright source curves around an enormous object, such as a galaxy, that is between the quasar and the earth. In 1979 a "Double Quasar" named Q0957 was discovered with the first confirmed case of gravitational lensing. This double quasar was caused by the refraction like effects of intense gravitational fields bending the light of a single quasar into two observed quasars (Bord, pg. 498). More than 30 such gravitational lenses are known today including the Einstein Cross (G2237 + 0305) where one quasar is refracted into four images of itself through gravitational lensing from a huge galaxy (Dekel, pg. 383).

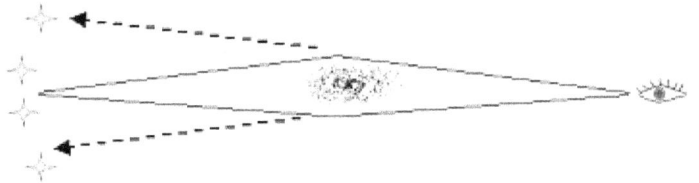

Gravitational Lensing

Gravity of the Matter

Gravitational Red Shift

The movement of the earth relative to a star or galaxy or the reverse, galaxy or star towards the earth produces a strange phenomenon known as Red Shift. Armand Fizeau, (inventor of the spinning toothed wheel experiment) described this spectacle in 1848 after noticing a shift towards red in the spectral lines of the light being viewed from stars moving away from the earth. Willem de Sitter had also noted this phenomenon in 1913 with the twin star Doppler Effect covered earlier. Red Shift can be defined as a stretching in the wavelength of electromagnetic energy detected by an observer or detector compared with the original wavelength emitted by the star moving away from the observer. This expanded wavelength is also directly proportional to the drop in the frequency of the radiated light. The opposite is also true with the compression in the wavelength of light coming from another object moving towards the observer and/ or detector, commonly called Blue Shift.

Edwin Hubble's (1898-1953) earth-shattering discovery in 1929 suggests that galactic Red Shift in light from stars and galaxies increase in proportion to their distance from earth, this is known as Hubble's Law (Ditchburn, pg. 340). Studies show that 38 of 40 galaxies observed are rapidly moving away from each other. This omni-directional galactic expansion shows that galaxies furthest away are moving the fastest away from earth– some at 57,000 kilometers per second or 13 million miles per hour – 2% the speed of light (Bord, pg. 220). This process implies that some sort of repulsive force exists between the galaxies, similar to an expanding gas or substance that is pushing galaxies outward and away from each other – somewhat like bread rising. Gravity in effect should be pulling galaxies together in what ultimately would be the "Big Crunch" . This possibly is why galaxies further away from each other are moving faster away from each other – gravitational forces are much weaker. Further research also shows that this expansion is accelerating additionally implying that an expansion agent also known as Dark Energy and Dark Matter (two different substances) exists in the universe. This substance is expanding between the galaxies and stars, forcing them outward, away from each other against their own gravitational attractions (Bord, pg. 493). Empty space may not be as empty as it may seem.

Gravity was now known not only to be able to bend light but even slow it down. A similar phenomenon known as gravitational Red Shift is noted in light emitted from very large stars, which is not related to movement but due to the immense gravitational fields of the star itself. The gravitational force of the star pulls at the light emitted from the star, slowing it down as if it had to climb up steep hill. This slowing effect on the light coming from the star causes the wavelengths of that light to become stretched, shifting the light toward the red end of the spectrum, called gravitational Red Shift. This Red Shift is only possible if the gravitational pull on the light emitted from the star is greater than the gravitational pull of the earth where the observer and detector are as the earth gravitational force has the opposite effect, speeding up the light (Ditchburn, pg. 315).

Gravitational Red Shift

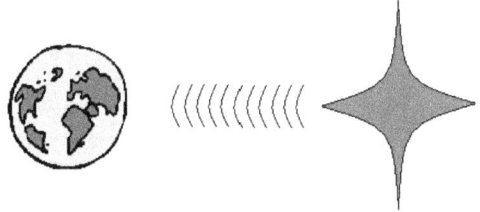

Gravitational Red Shift

Light reflected off the moon however, has the opposite effect, shifting towards the blue end, known as gravitational Blue Shift. This is only possible if the gravitational pull on the light reflected off the moon or any other object is less than the gravitational pull of the earth where the observer and detector are (Isaacson, pg. 148).

Gravity of the Matter

Gravitational Blue Shift

Highly accurate clocks onboard 24 GPS satellites circling the earth at 20,000 km (12,427 miles) altitude also are affected by Red Shift causing another phenomenon believed to be time dilation, a warping of space and time as predicted by special relativity (Bord, pg. 13). Gravitational Red Shift has a positive effect with GPS due to the weaker gravitational field of the satellite relative to the earth's surface. Time dilation however, has a negative effect due to the velocity of the orbiting satellite of 4 kilometer per second or 8,948 miles an hour relative to a stationary point on the earth, both tending to cancel each other out.

These GPS clocks must be synchronized to be within 50 billionths of a second in order to be capable of measuring positions on the ground within 4.5 meters (15 yards). These clocks must be corrected for "relativistic anomalies" due to gravitational Red Shift, otherwise they would run faster by more than 40,000 billionths of a second each day, changing the calculations of positions on the ground. Without these corrections, the clocks onboard the 24 satellites would be out to sync within a minute and a half – a matter of national security (Bartusiak, pg. 55). Einstein predicted that clocks would run slower in more intense gravitational fields (Isaacson, pg. 148). It is the earth's gravitational field that affects these clocks; therefore clocks at different altitudes should run at different rates having their frequencies shifted (gravitational redshift) due to gravitational differences. Two identical clocks at different altitudes will therefore run at different rates, with large enough difference to cause a variation in distance on the ground by as much as 13 km in a day.

Precession of Mercury

Another orbital phenomenon is the orbital precession of the planet Mercury. Mercury's orbit is an extremely elongated elliptical orbit approximately 36 million miles from the Sun that precesses slightly each orbit. Newtonian formulas explained this phenomenon by taking into effect the gravitational forces of all the other planets on Mercury including the effect from the Sun and its own rotation. Using these calculations, scientists calculated that Mercury's perihelion (the closet point of its orbit to the Sun) should shift or precess about 531 arc seconds per century. In reality, Mercury precesses about 574 arc seconds per century, 43 arc seconds more than the Newtonian predictions (Hobson, pg. 233).

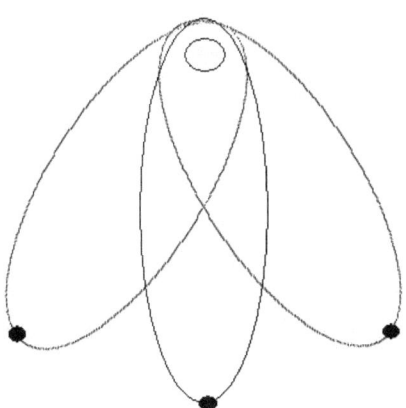

Mercury's Perihelion Shift

Gravity of the Matter

Newtonian formulas could not account for this change in motion. Something invisible was shifting the orbit ever so slightly, orbit by orbit. It has been suggested by some scientists that another planet, some call it Vulcan, may be in an orbit slightly closer to the Sun than Mercury, producing an additional gravitational force off center with the suns gravitational grip that could account for this discrepancy. Some other scientists have also theorized that there must be a certain amount of dust between the Sun and Mercury, which puts a drag on its motion and changes its orbit. No evidence however of dust or another planet has ever been found in this region. Einstein in 1915, calculated that the orbit of Mercury should precess by about 43 seconds of arc per century using the General Theory of Relativity, exactly what it does (Bartusiak, pg. 42).

Mercury's orbit ranges from 46 to 70 million km from the sun due to its elliptical orbit. Mercury is the fastest-moving planet in our solar system with a mean orbital velocity is 47.87 km per second with a minimum velocity of 38.86 km per second and a maximum velocity of 58.98 km per second, twice that of the earth. As Mercury changes velocity and enters into a stronger gravitational field during its Perihelion or closest encounter with the sun, its mass increases and its local time changes ever so slightly. These slight changes in mass and time due to the General Theory of Relativity help account for the missing 43 seconds of arc per century.

Tracking Venus and Mercury

If gravity of a large body in space such as a star could produce time dilation or a warping of space-time bending light rays, theoretically the sun could also produce this phenomenon. Working with astronomer Arthur Eddington in 1919 while viewing a solar eclipse, Einstein showed that the gravitational field of the sun actually did bend rays of light from a star. The position of the star was known but as its light rays passed close to its atmosphere, a form of Gravitational Redshift occurred, shifting its apparent position (Isaacson, pg. 257). The question arose that if the suns strong gravitational field affected light, could this same phenomenon be also evident with other electromagnetic energies?

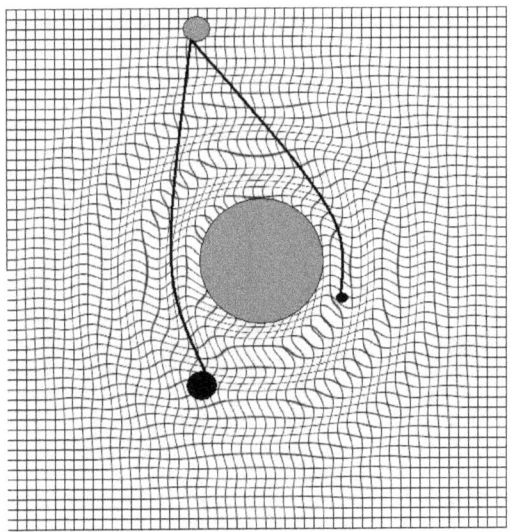

Radar Tracking of Mercury and Venus

In experiments carried out over several years beginning in 1964, Harvard Smithsonian astronomer Irwin Shapiro discovered a new variation of Gravitational Red Shift. Using MIT's newly built Haystack Observatory located in Westford, MA, Shapiro started tracking and recording Mercury and Venus's positions when earth and each of these planets were on opposite sides of the sun. By bouncing high-frequency radar signals off the planets, Shapiro found that as the planets started to pass behind the sun, the return time of the radar signals increased by about one five-thousandth of a second. This lengthened the 23-minute round-trip time to Mercury and the 30 minute round trip time to Venus of these radar signals, simulating an increase in the distance of the planets orbits as if the radar beam had bent or "dipped" changing course, increasing the distance it traveled to and from the planets (Bartusiak, pg. 51).

Cosmic Background Radiation

In 1964 while working on ultra-sensitive cryogenic microwave receivers, Bell Laboratory researchers Arno Penzias and Robert Wilson came across an unexplained microwave radio noise (Bartusiak, pg. 63). This pesky radio noise was isotopic, meaning it came from all directions, far weaker than the radiation given off by the Milky Way. They first assumed that their instrument was picking up interference by local radio and TV station noise from New York City but noticed that it came from all directions averaging in the microwave range of approximately one millimeter. The next assumption was that a pigeon dropping in the radio antenna's large horn was causing the microwave radio noise. After cleaning up the pigeon mess, the noise still frustratingly remained (Bord, 2000, pg. 304). Further research showed that this radiation was from so far away that it must have been generated over 15 billion years ago, a half a million years after the "Big Bang", before the formation of the planets when the universe was filled with a red-hot mist of plasma.

Optical radiation began to be generated as the first atoms began to form; the "primordial fog" was then lifted and the universe became relatively transparent, beaming with light (Bartusiak, pg. 226). Prior to the formation of atoms, the universe was probably amassed with only the building blocks of atoms: protons, electrons and neutrinos. Research on this newly found phenomenon shows that galaxies are all rushing away from each other at very high velocities and seems to show that the universe is constantly expanding (Ditchburn, pg. 340). It is believed that as the universe expanded, the hot plasma evenly distributed through the universe cooled. The light that was being generated at that time, mainly in the optical and infrared spectrum, became stretched, increasing its wavelength and lowering its frequency down from the optical range, down into the microwave range. While viewing the universe today with a traditional optical telescope, the background of space is now totally black. With a radio telescope however, there still remains a weak glow in all directions in the microwave radio frequency range of 160.2 GHz; this microwave radiation now being detected is scientifically known as Cosmic Microwave Background Radiation or CMBR.

Could this energy at the edge of our known universe be the remnants of the plasma that was generated during the Big Bang? Could this background noise actually be the energy of the material that all particles condensed from? If so, this relic that we see is the way it was 15 billion ago, which has by now has condensed into stars and galaxies, a universe we can never really see.

Dark Energy and Dark Matter

As previously mentioned, Dark Energy, the hypothetical expansion agent, possibly causes the rapid dispersion of our universe. As particles of matter gather great distance between themselves, gravitational attraction between them begins to wane, allowing this expansion agent to more rapidly separate particles from each other, accelerating them exponentially (Hawking, pg.48). The universe is believed to be composed of 73% of this mysterious Dark Energy, 23% of an unknown type of Dark Matter, and 4% of ordinary matter. Through understanding "Hubble's Constant" and calculating the known visible mass of galaxies, the universe should be expanding at a specific rate, but it is not. Through measurements of redshift in the light emitted from these galaxies, the universe is expanding much faster than first believed. The only way to explain this is to calculate in the missing mass (Glendenning, pg. 23).

The measurement from the Wilkinson Microwave Anisotropy Probe (WMAP), which was launched in 2001, has played a primary in developing today's Standard Model of Cosmology. WMAP, through the study of Cosmic Background Radiation has also determined that the cosmos is relatively flat and 13.7 billion years old (Glendenning, pg. 64).

This Background Radiation was determined to be homogenous or proportionally the same throughout the universe as a "smooth sea" of energy even though the stars and galaxies are clumped together in large masses with "nothing" in between them (Bartusiak, pg. 63). This energy seems to act as a hot gas, expanding outward, pushing everything around it away. As larger distances separate galaxies, the gravitational pull that attracts them diminishes, allowing this expanding "gas" more leverage, accelerating these masses faster and further apart. This was Einstein's Cosmological Constant which represented a repulsive force in the universe, counteracting gravitational forces that attract bodies throughout the universe (Isaacson, pg. 353). This constant was developed in order to maintain a balance in the universe, preventing it from collapse – a static view. After Hubble's discovery that indicated an expanding universe, this equation was modified, allowing for Hubble's Law (Ditchburn, pg. 340).

In the 1930's, Cal Tech professor Fritz Zwicky was first to conceive of Dark Matter in an effort to account for the high velocity of stars in the outer fringes of our own galaxy. Theoretically, these stars would spin off into space due to centrifugal force, but an unseen mass needed to be added, as gravity due to the visible mass of luminous stars was not enough. Dark Matter can also be used to account for gravitational lensing that curves light rays passing through the immense gravitational fields of galaxies. There has to be much more mass exerting gravitational forces than was seen visibly. The majority of Dark Matter is believed to be non-baryon, meaning that it does not have the atomic structure of visible mass such as the hydrogen atom and may not be able to interact electromagnetically with discernible mass other than gravitational forces (Glendenning, pg. 49). Dark Matter would evidently not be associated with large black holes in much of the universe, as gravitational lensing would indicate their presence, however, the existence of miniature black holes permeating the galaxies proportionately dispersed, is still a possibility.

Gravity of the Matter

A Challenge to Einstein

At first glance one might want to challenge the Michelson Morley Experiment by pointing out that the experiment would not work. It would not work simply for the fact that one beam crosses the aether wind at 90^0 and the other beam flows directly into the aether wind. It is being slowed down at first, but after being reflected off the mirror is sped up while flowing with the aether for a null effect.

The Michelson Morley Experiment

Let us use the flights of some aircraft for an example to emulate this experiment and compare two flight plans, both 2,500 miles one way. For the first aircraft, let's assume that a flight from Miami, Florida to Goose Bay, Newfoundland, Canada is both 2,500 miles one way and directly north in heading. For the second aircraft, the flight is also 2,500 miles from Boston, Maine to Seattle, Washington and is directly west in heading. The wind is straight out of the west at 100 knots at altitude. Both aircraft will maintain 500 knots airspeed the whole flight at the same altitude.

Using our aircraft navigation computers we take off with the two aircraft at the same time and fly round trip from Miami to Goose Bay and back and from Boston to Seattle and back, at the same air speed and wind conditions, timing both aircrafts round trip flights.

Flight Path

A Challenge to Einstein

The aircraft flying to Goose Bay averages 510 knots ground speed going to Goose Bay and 510 knots ground speed back to Miami with a 90^0 crosswind component both ways. It took 9.804 hours round trip to complete the flight. The second aircraft had to first fly into the wind and then return with the wind. Its first leg to Seattle took 6.25 hours while it fought the wind; the second flight back to Boston, however, only took 4.166 hours flying with the wind. The difference might surprise you; round trip for this second flight was 10.4166 hours, 37 minutes longer than the north south flight. Why the difference? Both flights had to fight the same wind and both flights were 5,000 miles round trip at the same airspeed! The reason? Diminishing returns! The harder the wind, the more time lost. At one point as you raise the bar, the wind will be at a higher velocity than the aircraft can fly and the aircraft will in fact fly backwards in reference to ground speed.

How does this relate to the Michelson Morley Experiment? The Michelson Morley Experiment was an experiment whose intent was to discover the "aether wind." As we can see in our aircraft experiment, there was a difference, but what if the wind was only one or two knots instead of one hundred knots? Would we have noticed the difference? How fast is the "aether wind"? Would the wind be one mile per hour? Would it be five thousand miles per hour? If this experiment would measure anything, how about pointing it up and down? Would it measure the effect of gravity on the light in the vertical plane? Isn't gravity, after all the "aether wind" anyway?

Gravity of the Matter

The "Big Bang"

Hubble's discovery in 1929 of an expanding universe lead scientist to try to determine what had happed in the past by picturing the universe going backwards in time, some 10 to 20 billion years ago, to a space-time "singularity" where everything was "one". All matter would come from a certain point where measurements were infinite, as if in a black hole (Asimov, Isaac, pg. 305).

This is how it may have started:

In the beginning, there may have been no such thing as time and space; then something happened. An immense release of energy occurred; a critical stage was reached, possibly similar to "critical mass" in a nuclear event out of the "singularity". A point was achieved in which the accumulation of mass and thermonuclear energy could no longer be sustained and a massive sub atomic nuclear like explosion occurred. In this huge explosion, existing mass is blasted into the smallest of particles as a hot plasma gas out into the far reaches of space. Huge amounts of energy are released as these smallest of particle ricochet off each other at enormous velocities, dissipating their energies out into the blackness of the universe.

Over time this energy continues to spread out from the center of the energy burst. As this outward energy dissipates into the regions beyond, a reversal of the flow or an "implosion" of these particles of mass begins to take place, back toward the center of the blast occurs. As time passes and as the particles energies and spins start to subside, higher energies and pressures now exist to the exterior of this spherical blast, pushing these small bits of matter together forming a denser pocket of matter in the low pressure center. Now cooler at the center, where the blast first took place, these "particles" now begin to go through a phase transition, reaching a *critical point* and precipitate or condense, forming a denser core – the beginnings of subatomic particles and atoms.

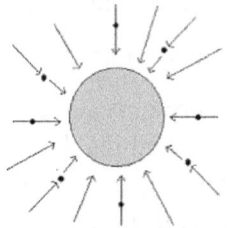

Condensation of Particles

One analogy that can be used is to look at these "particles" as low-pressure systems with pressure gradients such as isobaric lines that slowly decrease in pressure from the outside in. Just as miniature hurricanes or "black holes" , material or energy in this material slowly spirals in towards the center of this blob of spinning mass.

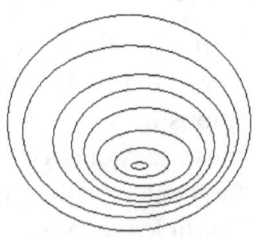

Pressure Gradients

The "Big Bang"

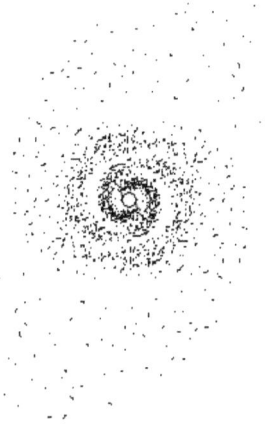

Low Pressure System

Over time, as energy disperses out into space, more and more of these "low pressure systems" begin to form. Elementary or fundamental particles flowing into these systems are spread out over great distances, spiraling in towards the center of these condensing storms. As these different systems come closer together they begin to react to each other even at great distances, as the space between them is not empty. Low-pressure systems spinning in the same direction repel each other even though far apart, bouncing off each other like tops spinning the same way. Systems spinning in opposite directions move together like gears in a clock, attracting one another and moving in unison.

Gravity of the Matter

Like Spins Repel

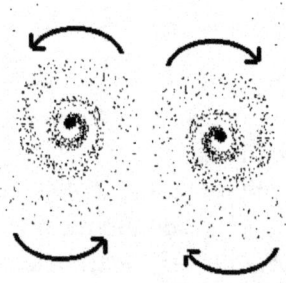

Opposite Spins Attract

As these systems or particles begin to interact with one another they transfer energy, bouncing off of, spinning with and spinning around each other until equilibrium of energy is attained. Various sizes of particles of different spins pair up with particles of opposite spins creating balance. Smaller particles of less mass with spins opposite of larger particles attract each other causing the smaller particles to begin to spin around the large particle, entering an orbit.

The orbiting particles become the life of these systems, speeding up and slowing down as they absorb or give off energy. The same number of particles in orbit balances the number of particles in the nucleus being orbited and an atom is formed.

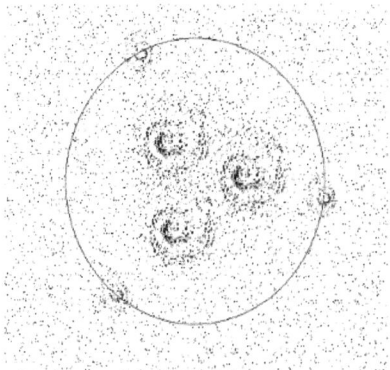

Particles Moving in Unison

The amount of energy given off or absorbed is directly proportional to the size of the particle and its velocity in orbit around the central system or nucleus. This amount of energy is a fixed value, a packet of energy, based on the velocity of the orbiting particle in the lowest possible orbit. This energy however, can be passed on to other particles at multiples of that value. As these orbiting particles absorb energy, they increase in momentum and move further away in their orbit around the nucleus. Also as these particles give off energy, they decrease in orbital momentum and move closer to the nucleus.

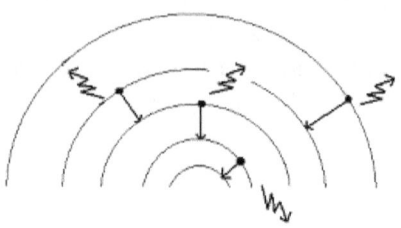

Release of Energy

As energy emitted from these systems is dispersed in these whole value "packets" of energy, they are likewise absorbed as whole packets of energy in other systems, even at some distance. The constant giving off and absorption of this energy makes the particles in orbit to move in and out in their orbits causing the nucleus to oscillate forming a wave pattern or "matter wave". This constant waving in and out of the orbiting particles disturbs the cloud of tiny particles that fills the space between these particles causing other orbiting particles even at great distances to resonate in unison. Particles in different orbits resonate at different wavelengths dissipating energy at various frequencies through the space around them.

Simply stated, the hypothesis stated here is that existing matter is made up of the smallest of particles that are gradually condensing as their energy dissipates out into the universe. These particles become the atoms, protons, neutrons and electrons that create all things in our universe. Energy is simply the movement of these particles in various ways and the interaction of these particles with each other even at great distances. The space between subatomic particles such as electrons and protons is not empty and the movement of one particle will ultimately move other particles transferring energy from one area to another at the speed of light even though far apart.

> *According to the general theory of relativity, space without aether is unthinkable; for in such space there not only would be no propagation of light but also no possibility of existence for standards of space and time.*

Albert Einstein (May 1920)

(Isaacson, Walter pg. 318)

Gravity of the Matter

Particles

As we saw previously, particles as we know of as electrons, protons and neutrons are possibly condensates of elementary or fundamental particles. These condensates may be formed by pressure gradients created by the cooling of a gaseous mass made up of these particles. During cooling, these condensates reach a saturation point and transition phase, bonding together at a specific energy level into a unified particle. Similar to the dew point, this may occur when specific pressures and temperatures are reached and "droplets" form. These droplets can only exist as a quantity of these elementary particles at specific energy states. These particles can be in a stable state as condensates of cooled fundamental particles formed into protons and electrons. They also may be manifested as moving corpuscle particles, for example, a photon that is produced by pressure waves created through compression.

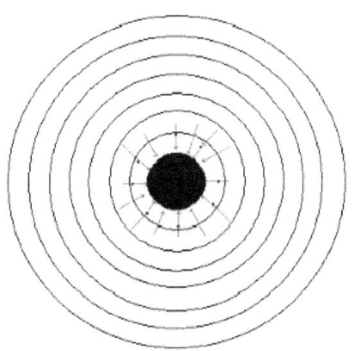

Condensates of Elementary Particles

These corpuscles or discrete particles are formed by pressure waves from energy traveling through an elastic *medium* of elementary particles. Again, a wave is defined as *a traveling disturbance of coordinated vibrations that transmit energy through a substance without net movement of that substance*. These discrete particles are created by energy passing through these relatively stationary particles.

These corpuscles are created by changes in the density of this *medium* from waves of particle motion in a specific direction. Since these particles can only exist under specific energy conditions, they remain confined to only regions of pressure or pressure gradients that create this phase change. In the form of an electron, this energy of moving waves of particles (spherical standing waves possibly) will be confined to a specific pressure level in the *medium* that exists as rings or shells around the nucleus of an atom. As energy is absorbed or emitted from these pressure gradients, existing electrons will dissipate into the *medium* of the region they are presently in and reform as a condensate of energy waves in particles of a different pressure gradient or shell surrounding the atom – disappearing and reappearing. Electrons may also be viewed as a sort of hybrid, a stable condensate particle yet wave-like in form. Energy given off or absorbed by these electron waves is only transferred in fixed "packets of energy", allowing these energies to only exist in specific shells specified by their momentum; the energy of the smallest shell or orbit possible, contains the energy of one of these "packets of energy". Each shell would then be created by multiples as if harmonics of this energy.

Electron as a Corpuscle of Energy

The energy that is being emitted or absorbed by these regions surrounding the atom is in the form of another discrete particle called the photon. This particle also exists as a pressure wave through the *medium* that can only exist under specific conditions as a "packet of energy". This packet of energy is given off by an electron as a pulse of energy in a given direction. The electron or wave of energy caught in the orbit of an atom that is emitting this energy then dissipates and reemerges as a wave of energy in a lower shell closer to the atom. The wave of energy given off from the electron as a pulse, then moves as a corpuscle of changing density in the *medium* of elementary particles in a straight line at the speed of light. This energy continues in that direction until it is absorbed by another electron in the orbit of another atom far away. Since energy cannot be created or destroyed and this particle can only exist under specific energy levels, energy is concentrated in one direction and cannot be dispersed. No "elementary particle" of this *medium* continues to move in this direction; they simply oscillate in place passing energy from one particle to another, compressing together into a denser mass in a chain reaction in one direction at the speed of light.

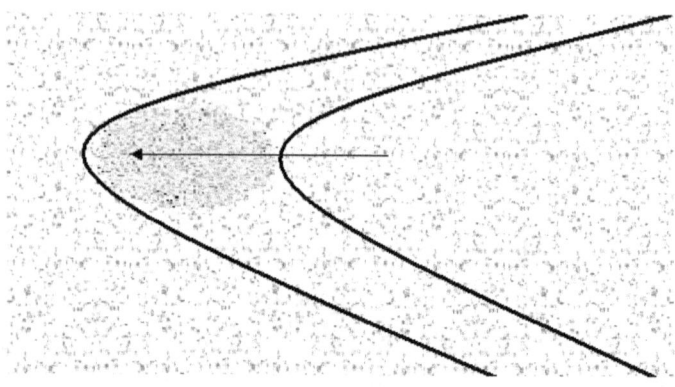

Photon as a Corpuscle of Energy

The photon as a particle, unlike the electron and proton, can only exist as a wave and has no mass of its own, just as a wave in the ocean cannot exist by itself. The photon and the wave on the ocean are simply the manifestation of energy – the movement of mass. The mass of the photon in a snapshot of time is the sum of "elementary particles of the medium" packed together in the wave front and may vary according to the density of the medium and the velocity of the wave through the medium. Its momentum or energy will vary as summations of "packets of energy". This process is continuously generated as electron waves surrounding atoms constantly absorb and emit this energy causing a back and forth waving of these particles in the *medium* of elementary particles. The different intervals or timing of these cycles depends on the energies and the type of atoms giving off this energy, producing different wavelengths of photon waves or colors of light detected in our eyes.

Particles

Electrons and photons do not only travel at high velocities in a given direction or orbit; they also oscillate back and forth at a given rate as they move in this direction. The energy of these electrons and photons can be mathematically represented in wavelengths. The lowest orbit possible would hold one packet of energy known as the Planck Constant and one full oscillation of the electron in its orbit would be explained as one full wavelength. An electron in higher orbits will have a higher a higher angular momentum, therefore a higher frequency of its oscillations and a shorter wavelength. The formula $p = h/\lambda$ can be used to show the momentum or energy of electrons in each orbit. Here p is the momentum, h is the multiple of Planck 's constant in whole numbers and λ, the Greek letter lambda, is the wavelength of the electron (Kafatos, Menas, pg. 33). The formula expressed as $\lambda = h/p$ shows the wavelength of the electron based on the multiple of the constant divided by the momentum of the electron (mass multiplied by velocity, mv). Energy released by the electron changes its momentum, thus its wavelength. This energy is transferred in another waveform out of the atom in the appearance of a photon of a specific wavelength and energy level.

Momentum or "p" of an electron in the <u>lowest possible orbit</u> is equal to the mass of an electron of $9.11*10^{-31}$ Kg multiplied by the velocity of the electron in this orbit of $2.19*10^6$ m/s (meters per second) which is calculated to be about $1.995*10^{-24}$ kilogram-meters per second. The wavelength or λ, can now be calculated with the energy of one Planck constant (h) to be 6.63 X 10^{-34} J-s (joule seconds).

Since λ = h/p, then:

λ = 6.63 X 10^{-34} J-s/1.995*10^{-24} kilogram-meters per sec.

or

λ = 3.32 X 10^{-10} meters or .332 nm (nanometers) the de Broglie wavelength discovered by **Louis de Broglie** in 1923 also known as the Matter Wave (Bord, pg. 385).

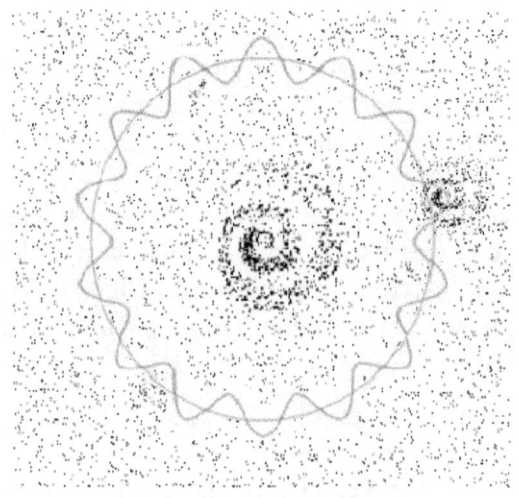

The Matter Wave

Particles

Particles whether they are waves of particles or particles as condensates are still created out of this medium and this medium is the foundation of mass and is mass. As particles approach each other, pressure gradients between the particles decrease, causing the differential in pressures around these particles, creating an invisible field that draws them together. Therefore photons and electrons will be affected by gravity and other types of fields created by these energies.

Attraction of Particles

Although all particles may be drawn together in this matter, as seen previously, particles with like spins repel and particles of opposite spins attract. Historically, three subatomic particles are explained in modern physics, electrons, protons and neutrons, in reality only two can exist on their own as stable particles, the proton and the electron since the neutron isolated alone will break down into a proton and electron within 15 minutes.

Studies have shown the masses of the particles to be:

Proton $=1.6726*10^{-27}$
Neutron $=1.6749*10^{-27}$
Electron $= 0.00091*10^{-27}$

What is interesting is that the mass of the electron and the proton added together is = $1.6735*10^{-27}$, very close to the mass of the neutron = $1.6749*10^{-27}$, a difference of $0.0014*10^{-27}$, which ends up to be an additional particle or particles with 1.53 times the size of an electron.

Under extreme conditions during the creation of our universe, high energy levels and atomic pressures turned excess protons in a nucleus into neutrons through the capture of electrons. These combinations of opposite spin particles, not unlike the hydrogen atom, created a neutral particle in the nucleus of the atom helping balance the strong repelling energies of proton, which have same spins and charges.

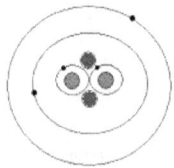

Neutrons in the Nucleus

In summary, a particle can also be viewed as a shockwave that is manifested as a larger single entity created by energy moving through a medium made of fundamental particles. The density and the energy (temperature) of these particles determine the velocity of this shockwave. This wave, similar to sound waves, can only exist at one velocity and cannot go any faster or slower in the material it is passing through. As the density or the temperature of this material changes, so will the shockwaves velocity.

The Photon

Light and other forms of electromagnetic radiation are believed by the modern scientific community to be the bombardment of particles called "photons" or what Einstein called "light quanta". These are particles that are believed to be ejected from the orbits of electrons, moving in a straight line and traveling at the speed of light until being absorbed by another object. Photons, having almost no mass, still have a noticeable displacement of material upon impact raising some questions. Einstein states that the faster a particle moves its mass increase until at the speed of light it has infinite mass. How then can anything travel the speed of light and have mass?

In the "Double Slit" experiment performed by Italian mathematician and physicist Francesco Maria Grimaldi (1816- 1863), a phenomenon called optical interference was created which can only occur if light is moving in waves (Ditchburn, pg. 119). In this experiment (also performed by Thomas Young), light from a single source is shone through a single slit in a barrier then continues through a second barrier with Double Slits. Light waves bend or diffract (breaking into pieces in Latin) as they pass through the slits causing the light waves to curve into what is called spherical waves. As light from both slits hits a viewing screen, the wave's peaks and valleys converge at the same points on the screen creating dark areas and light areas of contrast called an interference pattern. Waves that are in phase create constructive interference and brighten the area in which they strike. Waves out of phase cause destructive interference and cancel each other out causing the darker areas on the screen. This optical interference can only occur with light that is correlated or coherent, coming from the same light source (Ditchburn, pg. 116). Monochromatic light, meaning "one color," is more suitable for this type of experiment as only one wavelength is produced creating a more vivid interference pattern; white light is a mixture of all wavelengths and produces a less visible pattern. Similar interference patterns can be also seen with the interferometer that has been adjusted so that the two light paths are of different lengths causing them to arrive at the target out of phase.

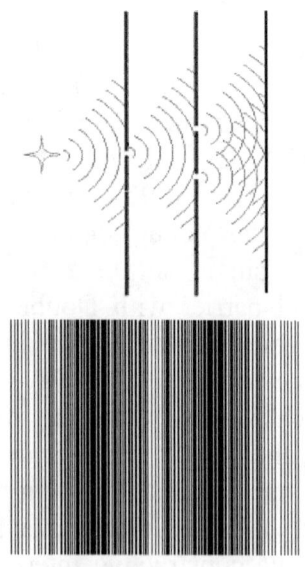

Double Slit Experiment

What causes these waves to diffract as they pass through the slits? This bending of light waves evidently is due to refraction from an interaction with the barrier as the light passes through the slits. The closer the light waves come to the barrier material, the sharper the bend.

The bending of light waves normally occurs when light moves from one medium into another and changes velocity, for example, light transitioning from air into glass. Light travels through air at approximately 185,226 miles per second, or 99% the speed of light in a vacuum. Light passing through glass however, only travels about 114, 989 miles per second, or 62% the speed of light. As the light transitions from air into glass, the light waves slow down and the wave front changes direction causing the light waves to bend or refract.

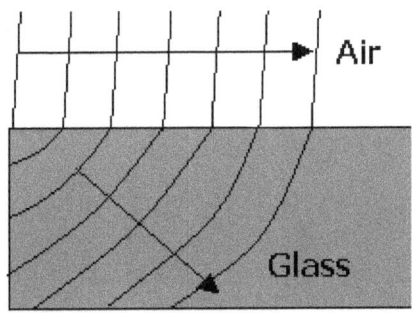

Refraction in Glass

This refraction can be seen with a spoon in a glass of water. The light that transitions from water into air after reflecting off the spoon speeds up and changes direction creating the illusion of a bent spoon. The refraction index varies from material to material based on the Optical Density of the material. The denser the optical density, the more the light bends as it passes from one material into another. Optical density is not the same as mass density, however there is a correlation but not linear.

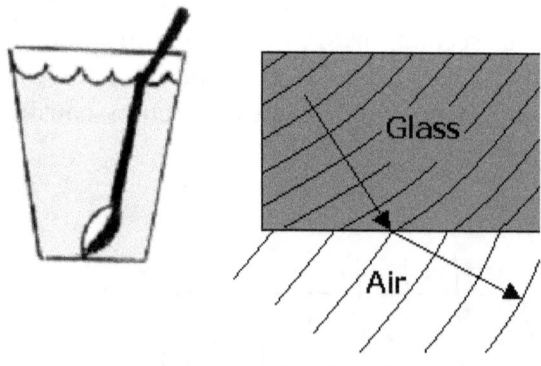

Refraction in Water

The light in the Double Slit experiment bends more the closer it is to the barrier material, as if passing through an invisible field of higher density material. As light passes through this area of higher density it slows down and changes its direction, bending outward. The combined wave becomes curved as a spherical wave; the two spherical waves created by the two slits then cause the interference patterns on the target surface.

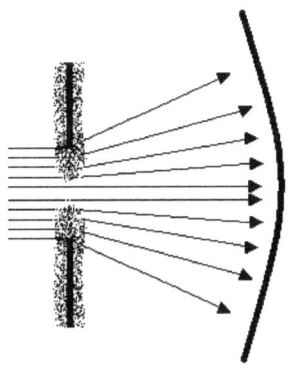

Slit Experiment Diffraction

In 1909 physicist Geoffrey Taylor performed a similar test using dark filters to block most of the photons, allowing only one photon at a time to pass through the slits. Over time the patterns of interference reappeared, showing that single photons will align themselves in a pattern of crests and troughs indicating that the single photon travels as a wave. Electrons also have been fired one at a time creating a similar pattern, refracting as it passes through the slit and falling on the target in crest and trough patterns (Ditchburn, pg. 643).

One at a Time

It is believed that the only way this could happen with single particles is if each photon or electron is divided and passes through both slits at the same time (Hawking, pg. 61). How else can crests and troughs be created? These patterns rely on interference, the interaction between two waves simultaneously, that create the constructive or destructive patterns on the target, light and dark spots. Otherwise an individual electron or photon would have to exist and then not exist; an on then off switching as it moves through space in order to produce this pattern by its self. When a photon strikes the target while turned "off" it would create a light spot; striking when "on" it creates a dark spot. Over time these will form interference patterns on the target. These light and dark spots stay in the same areas, and although created by moving particles, create a **standing wave** where the visible wave seems to be stationary.

One Photon Two Slits

Let's assume again that a medium exists as if a field around all particles and that this medium moves into these particles through a cooling or condensation process creating varying densities in the medium. Matter attracts matter; therefore, the medium closest to a large mass such as the sun will be denser than the medium out in empty space. As light energy is passed from one particle of the medium to another in waves of various wavelengths, it refracts slightly as it transitions through denser medium, changing the direction it is traveling. Further away in the less dense medium, light waves travel in a relatively straight line. In the Double Slit experiment, light waves passing closest to the walls of the slit curved as they passed through slightly denser medium creating a spherical wave moving towards the experiments target. These spherical waves from both slits created the interference on the target.

Einstein saw a similar event as light waves bend as they pass through the suns strong gravitational field changing the apparent position of a star. This proved to him that light was composed of matter – a particle that is now called the photon. Was this gravity that bent the light waves or the change in the aether's density close to the sun? Possibly they are the same thing.

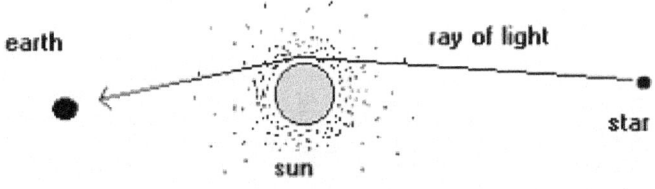

Light Rays

The flow of electrons also affects this medium that surrounds all things, generating electromagnetic waves in various wavelengths. The electron transfers electrical energy at the speed of light but dos not travel at the speed of light its self; much like air or water molecules that oscillate, moving sound at the speed of sound in the form of waves do not themselves move at the speed of sound.

Energy transferred from one electron to another in a wire carrying alternating current moves back and forth at the speed of light at a specific rate (the electrons themselves remain localized in the wire). The field surrounding the moving electrons generates a disturbance in the form of a wave that oscillates back and forth outside of the wire and is dispersed out into space in all directions as electromagnetic energy known as radio waves. This wave must have something waving much like air or water particles do.

Electron in a Wire

The Photon

The electrons themselves can stay relatively confined and only pass the energy from one electron to the other at the speed of light. Electrons generate electromagnetic radiation by jumping "back and forth" from one shell of an atom to another while orbiting around atoms or when jumping from one atom to another. Notice the "back and forth" motion? Remember, the "five ball pendulum", where there were five or so balls hanging on strings lined up in a row next to each other? Energy is transferred from one ball to another with very little movement of each ball; only the first and last balls really move at all. The same is true with electrons and the transfer of their energy and may also be true for energy traveling as light waves through this medium that envelopes all things.

Five Ball Pendulum

Wave patterns not only are evident in the fore and aft movement in the direction of propagation, but in all directions, which bring us to the phenomenon of polarization. All particles find an equilibrium or neutral state when they expel all energy and come to rest. As energy passes through these particles however, they oscillate back and forth like pendulums, up and down, left and right or fore and aft. This back and forth waving of energy is another indicator that light moves in waves.

Pendulum

In the electromagnetic realm, this direction of wave motion is called polarity and is evident in all wavelengths of electromagnetic energy. The use of polarized sunglasses and the vertical or horizontal positioning of radio antennas are all examples of common utilization of the polarity phenomenon.

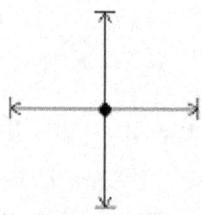

Wave Dimension

For example as light waves travel through certain crystals such as tourmaline, the crystal formations form tunnels that only certain wavelengths of light with a specific polarity will pass (Polkinghorne, pg. 18).

The Photon

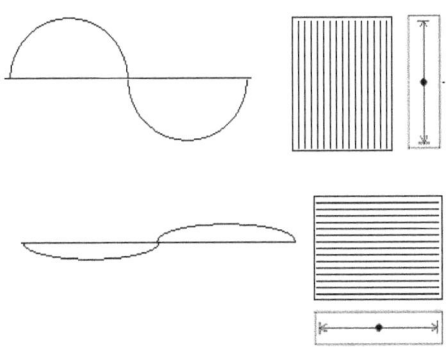

Vertical and Horizontal Polarity

Waves with opposite polarities become blocked such as vertically polarized waves trying to pass through a horizontally polarized pane of glass.

Opposite Polarity

All materials molecularly are constructed differently and have various responses to the electromagnetic energy. Some reflect, some absorb and some pass the energy straight through them. When absorbed, most of the radiation is then re-radiated back out at other wavelengths such as infrared radiation.

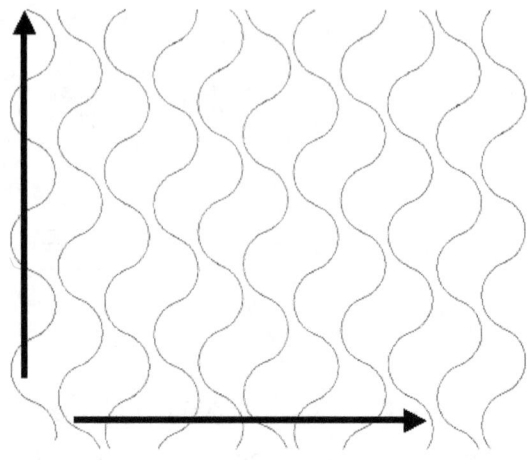

Electromagnetic Waves Disbursement

Particles move back to their natural state passing energy to the next particle in waves called plane waves or linear waves. Some waves not only move energy in the direction of the wave but also perpendicular to the wave in what is known as a transverse wave – these types of waves are only possible in **solids**. Energy in the electromagnetic realm moves at the speed of light in all directions.

Waves or Particles

In the year 1900, as previously discussed, German physicist Max Planck hypothesized that energy emitted or absorbed by an atom can only be transferred in distinct quantities, which he called "quanta", Latin for "how much". This "quanta" soon played a critical role in the new study of quantum mechanics as it required that oscillating atoms only possess the energy values of E or a whole number multiples of that value. The formula for E = h X f, where Planck's constant "h" is multiplied by the frequency "f", of the oscillating body (Bord, pg. 374).

The rationale behind this formula is that one full oscillation of a particle; let's say an electron, during a specific time frame, would use a specific amount of energy. Doubling the oscillations during the same time period would take twice the amount of energy and tripling the oscillations would therefore take three times the energy and so on. This oscillation is the frequency at which the electron vibrates and as it turns out is directly proportional to wavelength of the light waves emitted from an oscillating electron, whether it is from a star or a light bulb.

Planck's constant, "h" is measured in Joules (symbol J), the International System of Units (SI) energy measurement for heat, electricity and work. Planck's constant, h = 6.63 X10^{-34} Joule seconds (Bord, pg. 375).

There are two oscillating bodies or particles that are commonly referred to as energy carriers in the quantum realm, the electron and the photon. Each of these particles can be scientifically considered as either particle or wave in nature as mathematical formulas have been developed that seem to indicate them as both. The neutron, outside of the nucleus has a very short life span of only a few minutes, breaking down, believe it or not into a proton and an electron in a radioactive decay process known as beta decay.

Even within the nucleus, this radioactive decay occurs naturally in some elements, as in carbon 14, where an electron is ejected from the nucleus at a high rate of speed and a neutron spontaneously converts into a proton; a smaller particle known as a neutrino is also emitted as a release of energy (Bord, 2000, pg. 417). Carbon 14's nucleus contains 6 protons and 8 neutrons and converts into Nitrogen 14 with 7 protons and 7 neutrons when the neutron goes through beta decay and becomes a proton.

Many other short-lived elementary particles have been discovered in recent years, bringing to light the complicated constructs of the subatomic world. These include leptons and quarks, the building blocks of electrons and protons, as well as antiparticles, which mimic other atomic particles inversely as if viewed in a mirror having opposite charges and spins. The contact of a particle and anti-particle results in the annihilation of both particles in a large burst of energy (Bord, 2000, pg. 457).

Elementary particles are the building blocks of the universe and are the fundamental constitutes of matter, either bosons or fermions depending on the type of spin (up or down). Particles normally linked with the construction of matter are called fermions and have a half-integer spin (odd integer (number) spin +1/2). Fermions are divided up into twelve categories called "flavors" and three other groups called "colors" (Bord, 2000, pg. 479). Particles associated with fundamental forces - the strong, weak and electromagnetic and possibly gravitational forces are called bosons, named for Satyendra Nath Bose, having an integer or whole number spin (Bord, 2000, pg. 458).

Fermions:

1) **Quarks** - up, down, charm, strange, top, bottom

2) **Leptons** - electron neutrino, electron, muon, muon-neutrino, tau and tau-neutrino

Bosons:

1) **Gauge bosons** - gluon, W and Z bosons and photon

2) **Higgs boson** and the theoretical **graviton**

Only four charged subatomic particles exist that are stable particles, the electron, proton, positron and antiproton (Asimov, Isaac, pg. 246). About 500 other particles have been identified in experiments using particle accelerators in recent years, all of which are unstable and exist only for short periods of time. Subatomic particles are composite particles created out of fundamental particles such as quarks and gluons that have no known substructure. These particles can only exist under certain conditions, known as "confinement" such as in the neutron, which decays into a proton and an electron in a matter of minutes outside of an atoms nucleus (Bord, pg. 417). Fundamental particles such as leptons and quarks are highly unstable and can only exist in confinement, attached in pairs or triplets by the "strong force"; these elementary particles in turn are the main components of the electron (a lepton), proton (two up quarks and one down quark) and of course the neutron.

Although these particles may seem to exist as individual masses, they actually may simply exist as waves in the field of fundamental particles that fill the vastness of empty space. As mentioned earlier, subatomic particles possibly only exist as a quantity of these smaller fundamental particles under certain conditions. As energy moves through these particles they temporally coalesce in a form of phase transition, passing energy from one particle to another in much the same way as do fans executing the "wave" in the stands of a sports event. Since energy cannot be created or destroyed (Law of the Conservation of Energy), this energy continues on in waves from particle to particle. This energy at the quantum level in the form of Plank's Constant also cannot be divided, transferring all of its energy to the next particles.

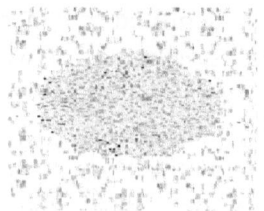

Electron as Particles Temporally Coalesced

Under specific conditions, this pressure wave possibly becomes trapped in a continuous circular motion, confined to a specific pressure gradient around an atom where it alone can exist, manifesting itself as an electron. As energy is absorbed into this wave, the wave changes momentum and instantaneously repositions itself in a pressure gradient where different conditions allow the "particle" to reappear as an electron in another shell or pressure gradient; such as transitioning from orbit 5 to 8 at a new velocity. Think of this as fans performing the wave around the stadium only in row 55, then halfway around stopping and shifting instantly to only row 75 for the rest of the way around. Although there are people in every seat, the wave manifests itself as people stand and raise their hands one after another and then sit down in a continuous fashion. The distance the wave has to travel as it circles the stadium varies from row to row as well time it takes to complete one circle.

Transition from Orbit 5 to 8

If we look at the electron as a single particle in orbit, as if a satellite around the atomic nuclei, the electrons wavelength of the smallest orbit of the hydrogen atom with a velocity of the $2.19*10^6$ m/s (meters per second) is: (λ is wavelength)

λ = h/mv (Planck's Constant and Momentum), then:

λ = 6.63 X 10^{-34} J-s/1.995*10^{-24} kilogram-meters per sec.

or:

λ = 3.32 X 10^{-10} meters or .332 nm (nanometers)

Nanometer = one billionth of a meter (0.000000001)

If r is the radius of an orbit then the circumference of that orbit is = $2\pi r$, which is also its wavelength.

Since $2\pi r$ = λ = .332 nm then the radius of the smallest orbit or orbit **1** is:

1 radius = $\dfrac{.332 \text{ nm}}{2\pi}$ = $\dfrac{.332 \text{ nm}}{2 \text{ X } 3.14}$ = $\dfrac{.332 \text{ nm}}{6.28}$ or .0529 nm

Orbit **2** would **theoretically** then be **2λ** or

= 2 X .332 nm = .664 nm.

The circumference of the next orbit or orbit **2** is:

2 radius = $\dfrac{.664 \text{ nm}}{2\pi}$ = $\dfrac{.664 \text{ nm}}{2 \text{ X } 3.14}$ = $\dfrac{.664 \text{ nm}}{6.28}$ or .106 nm

Orbit **3** would then be **3λ** or = 3 X .332 nm = .996 nm.

The circumference of the next orbit or orbit **2** is:

3 radius = $\dfrac{.996 \text{ nm}}{2\pi}$ = $\dfrac{.664 \text{ nm}}{2 \text{ X } 3.14}$ = $\dfrac{.996 \text{ nm}}{6.28}$ or .159 nm

Waves or Particles

Orbit 1 radius = 1λ = .0529 nm
Orbit 2 radius = 2λ = .1057 nm which is 0.0529 nm further
Orbit 3 radius = 3λ = .1586 nm which is 0.0529 nm further
Orbit 4 radius = 4λ = .2115 nm which is 0.0529 nm further
Orbit 5 radius = 5λ = .2643 nm which is 0.0529 nm further

Energy = h (Planck's Constant) X Frequency

Frequency = $\dfrac{\text{Speed of light}}{\text{Wavelength}}$

Wavelength = $\dfrac{\text{Speed of light}}{\text{Frequency}}$

Speed of light = c Frequency = f   Wavelength = λ

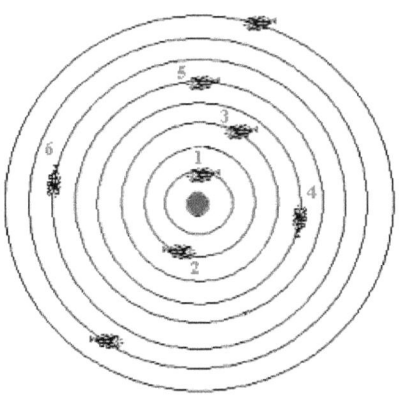

Waves of Particles

We see that if we increase each orbit by multiples of the wavelengths, each orbital radius is increased proportionately as in the previous illustration, 0.0529 nm further out for each new orbit. This is not the case however; each orbit changes according to variations in energy in fixed quantities of multiples of h, Planck's constant, not fixed distances from the nucleus. These orbits are not measured in radius or circumference, but in Electron Volts. Each orbits radius decreases slightly the further out in orbit the electrons are. As an electron transitions from one orbital or shell to another, the distance between orbits also changes.

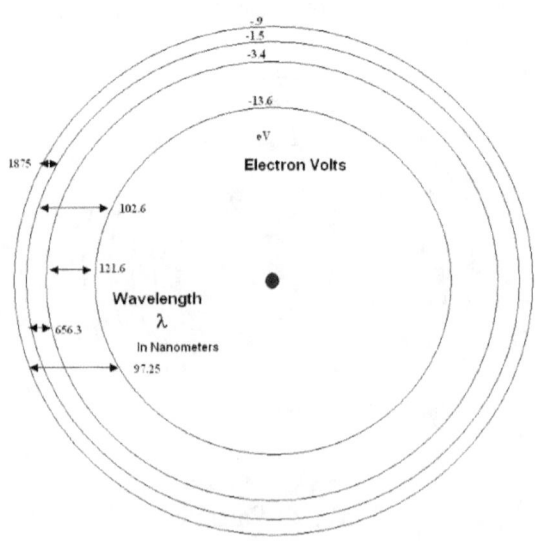

Radius Change in Orbit

The difference between orbit 1 and orbit 3 is greater than the difference between orbit 4 and orbit 6. The velocity of the electron also varies from orbit to orbit. As the distance of the electron from the nucleus increases, the force that pulls the electron towards the nucleus decreases by the square of the distance. This phenomenon is common to many processes such as the change in gravity, magnetism and the change in intensity of electromagnetic energy or light as the distance from the source increases. This in turn has a direct effect on the electrons velocity.

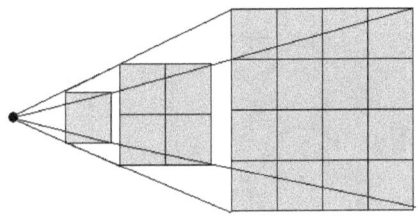

Attraction Decreases by the Square of the Distance

If an electron has the mass m, and the nucleus has mass M, and the electrons distance from the center of the nucleus is r, then g is the force that the nucleus exerts on the electron or g = $Gm\ M\ /r^2\ /m$ (where G is a universal constant for the attraction of mass). If the electron would fall towards the nucleus, the nucleus's attraction will cause it to accelerate toward the center of the nucleus.

According to Newton's second law (F = ma), this acceleration g must equal:

$g = (Gm\, M/r^2)/m$, or $g = GM/r^2$

For the electron as a single particle, the proportionate change in energy input does not produce a proportional change in orbit. The velocity of the electron varies according to the change in the nucleus's attraction due to the change of distance in radius from the nucleus.

The **angular momentum** (p) of each orbit or mrv is:

$$F = \frac{mv^2}{r} \qquad \frac{v^2}{r} = \frac{GM}{r^2}, \quad \text{or} \quad v = \sqrt{\frac{GM}{r}}$$

(**M**ass x **R**adius x **V**elocity)

$$mrv = \frac{h}{2\pi}$$

Looking at the electron as a particle, we see that at the same level of energy, $2h$, we can have different velocities and different linear momentums. As far as the electron goes, the higher its orbit, the lower its velocity and the lower its linear momentum. Likewise, the lower the orbit of the electron, the higher its velocity and the higher its linear momentum. Angular Momentum (mrv) must therefore be calculated in order to show the total energy of the electron.

We also see that attraction forces change according to variations in orbits; the higher the orbit, the lower the attraction forces are. This change in distance has a direct effect on the velocity and linear momentum of an electron. Also, the higher the energy of the electron, the further out in orbit an electron will go; but the further out the electron is in orbit, the slower it travels and the less linear momentum it has. What then creates the fixed wavelengths of energy emitted from these systems? How do these slower electrons create the higher energy photons of shorter wavelengths? It seems the longer wavelength of the electron, the shorter the wavelength of the photon it creates – a paradox of sorts.

When sound is created from a guitar string, the string oscillates back and forth at a specific rate every second. Its wavelength is equal to the length the string physically moves, first one direction and then the other. The longer the wavelength of the strings oscillations, the lower the frequency and the less energy used; the shorter the wavelength, the higher the frequency and the more energy used. To double the frequency of the strings oscillation is to double the energy radiated by the string at the same amplitude. This strings motion disturbs the air around it and the energy it radiates is dispersed into the air as compression waves or plane waves. Each of these plane waves travels away from the string at the speed of sound in that *medium* (air) at fixed wavelengths.

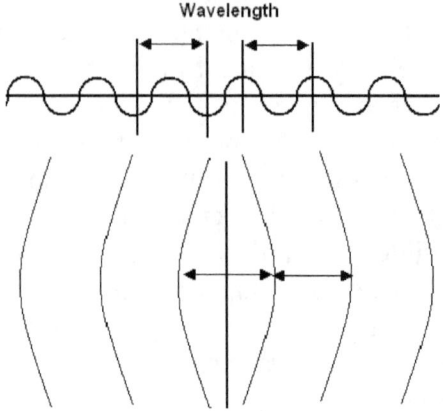

Wavelength

Guitar String Oscillations

An electron likewise oscillates back and forth as a guitar string. It does this in two ways, during its normal orbit around the nucleus of an atom in a circular motion at multiple levels and as it momentarily jumps from one orbit to another. The circular motion of the electron in orbit around the nucleus traveling a full 360 degrees is one full wavelength with each change in orbital radius a different wavelength. The velocity of the electron in a specific orbit produces a repetitive motion that repeats thousands of times per second, measured in Hertz (Hz). The faster the electron oscillates, and the smaller its orbit, the higher the frequency of the electron. The larger the orbit, the slower its velocity and the lower the frequency of the electron. The wavelength of the electron is dependent on its velocity and orbital circumference, not just the diameter of its orbit.

104

One thing to keep in mind is that although an electron in a higher orbit is slower than lower orbiting electrons, it has more energy due to angular momentum. Faster electrons in lower orbit likewise have less energy.

An electron also is constantly changing orbit as it absorbs energy bombarding it from all directions by photons. Each time the electron absorbs energy from a photon, it flies further out in orbit, and then releases the energy it absorbed and retreats back to a stable orbit. Each of these jumps or quantum leaps of the electron absorbs or emits a photon in fixed energy bundles previously discussed.

What the different here is, that unlike the guitar strings wavelength, which is proportionate with the strings vibrations, the energy released from an electron works in an inverse way. The larger the movement of the electron, whether in orbit or quantum jump, the shorter the wavelength of the energy emitted. Likewise, the shorter the jump or smaller the orbit, the longer the wavelength of energy released.

As covered earlier, a particle can possibly be viewed as a shockwave whose velocity is limited by the density and temperature of the material or *medium* the electron wave is flowing through. When viewing an electron in this matter, the closer to the nucleus an electron wave moves, the denser and cooler the *medium* is that the electron wave is moving through (since the particle itself may be a condensate of these cooled particles in wave form).

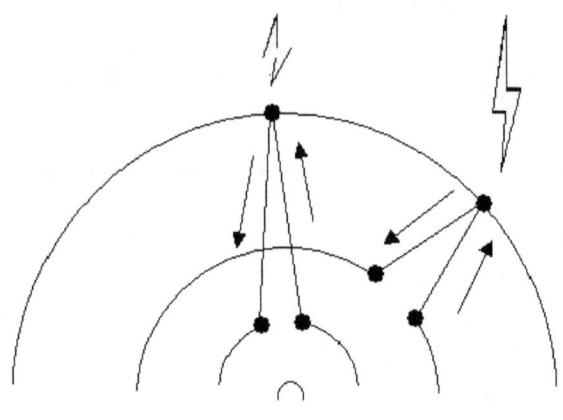

Quantum Leap

   The further out that the electron wave moves from the nucleus, the hotter and less dense the *medium* in which the wave is moving through theoretically becomes.   The electron as a fixed wave bundle of energy can only exist as a shock wave in one region at a time, at a specific velocity. If energy were added as a second fixed bundle of energy to the first wave (another photon wave is merged with the electron wave), the momentum would then be doubled (2*h*). With the momentum doubled but the mass of the combined particles in the wave still the same, the velocity and the orbit of the electron therefore must change. However, the present position of the wave in the *medium* it is presently moving through can no longer exist at this higher energy level, forcing the wave pattern to reform as a shockwave further away from the nucleus in a less dense region. This new wave then travels in a circular pattern around the atom further out staying within the new density bounds at a different velocity. This wave energy will be created at a higher angular velocity (mvr), which is actually a lower linear velocity but at a higher energy level. We must remember here that the electron itself is not a particle moving the distance of low orbit to high orbit or vice versa, it is simply energy dissipating in one level and instantaneously reforming in another level.

We can again picture this as fans doing the wave around a stadium only in row 55. The more people in the row generating the wave, the longer it takes for the wave to move around the stadium. Inner rows have less distance to travel around the stadium but may have more people packed in the seats, slowing down the wave. Seats near the back of the stadium have many more seats for the wave to move through but possible has less people in the seats changing the velocity of the wave as well.

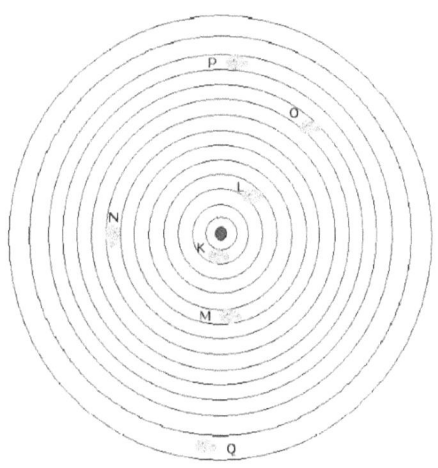

Gravity of the Matter

	Temp in Kelvin	Velocity of Light
K	10.13	2,189,856
L	20.46	4,421,846
M	41.32	8,928,773
N	83.44	18,029,345
O	168.49	36,405,593
P	340.21	73,511,667
Q	686.97	148,437,772
∞	1387.16	**299,731,634**

    The previous illustration shows the seven shells that an electron wave can move through. The illustration shows how a possible temperature and density change of a *medium* from one shell to another affects the velocity of a shock wave through that *medium*. The velocity above Q, the highest orbit, breaks the bond of the nucleus at the speed of light c, 300,000 km/sec, an escape velocity if you will, sending energy out through open space in the form of a photon in a straight line.

Fields and Forces

There are four fundamental interactive forces in nature. These four categories of forces studied in quantum mechanics are gravitational force, electromagnetic force, the weak force and the strong nuclear force; the weakest of these is gravity.

Gravitational force is the affect that every particle has on other particles and its force is directly proportional to the combined mass of objects within the field; this field's strength decreases by the square of the distance between the particles. For example, the sun and the earth: here the total gravitational attraction the sun has on the earth must be added to the gravitational pull the earth has on the sun. It is not just the powerful gravitational pull the sun alone that pulls on the earth but also the earth's own pull on the sun. Likewise, a man standing on the earth not only is attracted by the earth's pull on him but also his mass that is also attracting the earth. The greater the total mass (earth + sun), the greater the gravitational force that is pulling the objects together.

Long ago, Aristotle had reasoned, supposedly erroneously, that the velocity of falling objects were dependent on their weights. Galileo a hundred of years later discovered that the velocity of two objects of different weights fell at the same rate; this view is still held today. In fact, Aristotle was somewhat correct. A marbles rate of acceleration towards the earth is not only caused by the earth's attraction of the marble but also the marbles attraction of the earth. The acceleration rate towards the earth is due to the combined mass of both objects; therefore, a bowling ball will have an ever so slightly higher rate of acceleration. This is due to the fact that the combined mass of the earth and the mass of the bowling ball is slightly greater than that of the combined mass on the earth and the marble.

Gravity today is believed to be manifested by a mass-less particle called a graviton. This is done through an exchange of gravitons in gravitational waves between the particles of two bodies. This force, unlike the other three forces, only attracts and does not repel. Although it is the weakest of the four forces, its effects can be generated over a very large distance (Hawking, pg. 72).

The **electromagnetic force** is a force that is both attractive and repelling in nature and only acts between two or more electrically charged particles. Here, like charges repel and opposites attract, positive repels positive but attracted negative. Examples are where a proton attracts an electron but repel another proton; this is the attracting force that causes an electron to orbit the nucleus of an atom that contains the protons (Hawking, pg. 73).

As protons and electrons particles interact with each other, orbiting electrons will in certain situations be shared with the nucleus of other atoms; in a sense, tying the two atoms together in what is called a **covalent bond.**

Fields and Forces

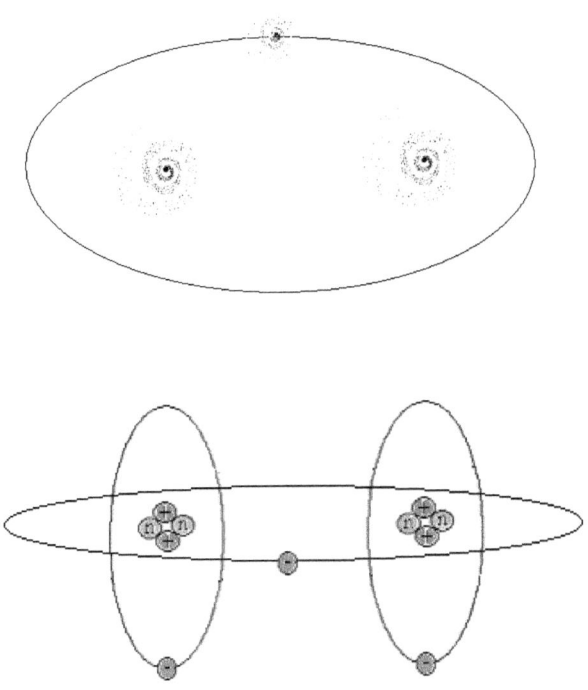

Bonding of Particles

In electrically balanced conditions, electrons and protons equal each other in numbers. In cases however where imbalances occur, electrons will try to passed from an area were excessive electrons are grouped to a region of electron deficiency. If a conductive path is available, a flow of electrons from one area to another in an electric current will occur– the principle of the battery. As these electrons move from one atom to another, a field is produced surrounding the electrons, moving with the electrical current disturbing space around them as they flow with what is commonly called an **electric field**.

Electron Flow

Magnetic fields are caused by the motion of electrons in a circuit as in the case of an electromagnet. Here the electrical field creates an attractive or repulsive force depending on the direction of the flow; its intensity is proportional to the amount of current in the circuit. A natural magnet such as magnetite also has a magnetic field but is caused by the alignment of atoms in the substance and possible flow of the electrons naturally within the substance. The synchronous movement of electrons moving in unison and perfect alignment of the electron orbits within the magnetic material produces these magnetic fields.

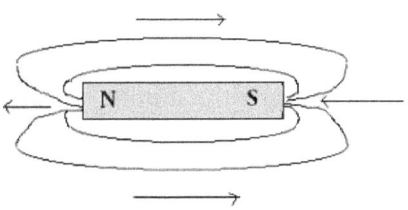

Magnet

The **electromagnetic field** in the form of a radio wave is also a variant of this electromagnetic force. Light waves are also electromagnetic waves that are at higher frequencies than radio waves and are manifested by the ejection of photons by the movements of electrons.

Radio Waves

The **weak nuclear force** is found in nuclear radioactive decay such as beta decay, the breakdown of the neutron into an electron and a proton. Here, outside of the confines of the atom, a neutron no longer remains a neutron and breaks down due to the exchange of heavy W and Z Bosons. Particles react differently at different levels of energy, at specific energy levels symmetry can be broken causing particles to change, spontaneously breaking down (Hawking, pg. 74).

The **strong nuclear force** is the force that holds the building blocks of particles together. The quarks that make up protons and neutrons and the nucleus of the atom are possibly held together by the gluon. This gluon binds them together in a process called confinement, combining only specific combinations of different types of quarks (green, blue or red). Combining anti quarks and quarks may create the unstable meson of which then breaks down into an electron. The high energies of particle accelerators weaken the strong force, allowing these particles to be broken down into what appears to be "free" quarks and gluons (Hawking, pg. 76).

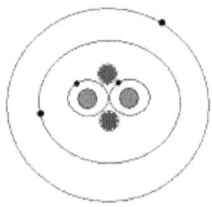

These forces are all forms of energy. Energy as we have learned is the movement or the potential movement of mass. Scientifically speaking, energy is a variable quantity that is a characteristic of an object. Energy is the ability to do work or that which is transferred when work is done (Bord, 2000, pg. G-1). There are several ways to describe energy, kinetic, potential, gravitational, electromagnetic, chemical and nuclear. Energy can be transformed from one form of energy to another such as light into heat, but cannot be created nor destroyed. This is known as the Law of the Conservation of Energy (Bord, 2000, pg. 98) written in Sir

Isaac Newton's *"Philosophiae Naturalis Principia Mathematica,"* Latin for "Mathematical Principles of Natural Philosophy"- one of the most influential scientific works ever published.

As earlier discussed, there must be a *medium* or substance that all of these different forces travel through. It is the means of which all wavelengths of the electromagnetic spectrum are transferred. It is the guide wave for the particle world; photons, electrons, neutrons, and protons are all influenced by its existence or exist, as waves of it and it may also be the pressure or force of attraction called gravity. If energy exists as waves in empty space, what is it that is waving? In the sea, it is water molecules, in the air, air molecules. All particles may influence or even be composed of this *medium*. As we saw before with the radio antenna, the motion of the electrons moving through the wire transferred energy into the *medium* and back from the *medium* into the receiving antenna even through "empty" space.

Gravity of the Matter

Gravity

What is gravity? What is it that causes the acceleration of objects as they fall? Galileo determined that all objects, not accounting for air resistance, fall at the same rate at 9.8 meters per second per second no matter what size or weight on the earth.

Einstein noticed the similarity between gravity and acceleration and developed the "principle of equivalence", asserting that gravity and acceleration are the same. In a thought experiment, Einstein would imagine a man closed up in a in a "spacious chest" resembling a room. This room would be as if it were, in orbit with no gravity, when suddenly the room is accelerated by some force at a constant rate of 9.8 meters per second/ second. In this situation, not being able to see outside, the man perceives that this force is gravity, making gravity and acceleration indistinguishable (Einstein, pg. 67).

If gravity and acceleration are the same, how can we account for this "acceleration" in our universe? As a car takes off from a standstill and accelerates, the driver feels forced back into the seat. The higher rate of acceleration, the more the driver is pressed back. If the car accelerates at 9.8 meters per second, per second (per sec^2), the driver would feel the equivalence of 1 G of pressure exerted into the back of his seat.

Now visualize the surface of the earth expanding outward at a rate of 9.8 meters per sec^2. All objects on the earth would maintain their weight similar to gravity. The problem with this however is that everything (buildings, houses, cars etc.) would also have to expand outward proportionally with the earth to stay in a balanced perspective and the rest of the universe likewise would have to expand at the same rate to stay in a proportional balance – everything would envelop everything else.

Now comes a dilemma. Other objects with different gravitational forces must also be taken into consideration; for example, the moon, which has one-sixth, the gravitational force of the earth. With the earth's surface expanding outwardly at 9.8 mps^2, the moon's surface would be expanding only at 1.6 mps^2. It wouldn't take very long at all to notice from an earth perspective that the moon would be shrinking very rapidly. In order to correct for this dilemma one would again have to change space and time to straighten out this dimensional warping of space. The rate of lunar expansion therefore may be perceived to have sped up in order to keep a universe in balance or perceived distances would have to be greater to have this perception of no change.

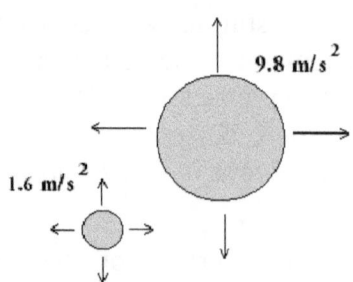

Different Rates of Expansion

Gravity

This seems ridiculous. Again, we would have to resort back to the Lorentz Contraction Theory of dilated space and time in order to correct distorted space. However you look at it, all mass would eventually envelope all mass in this scenario, and everything would be annihilated; the earth would swallow up the moon and the sun, likewise the earth.

$$l'=l\ \sqrt{1-\frac{v^2}{c^2}}$$

Dilation of space

$$t'=\frac{t}{\sqrt{1-\frac{v^2}{c^2}}}$$

Dilation of time

Let's now look at another possible answer to this problem. If this gravitational force of acceleration is not caused by rapidly expanding mass, what else could it be?

When a boat crosses a rapidly flowing river, the water current carries it along causing it to drift down stream. The same could be happening here. A rapid cooling of the universe after the big bang could be causing the condensation and rapid contraction of this medium, the aether cloud that surrounds all matter. As this material condenses, it forms in various levels of pressure, creating a "flow" of subatomic particles moving through and around all mass in its vicinity. This material would create regions of varying pressures; as the materials cool and condense the pressure drops causing higher pressures and energies to exist further away from the condensing mass. The higher pressures outside of the condensing mass and lower pressures inside mass pull mass in towards other mass, similar to wind on a sail.

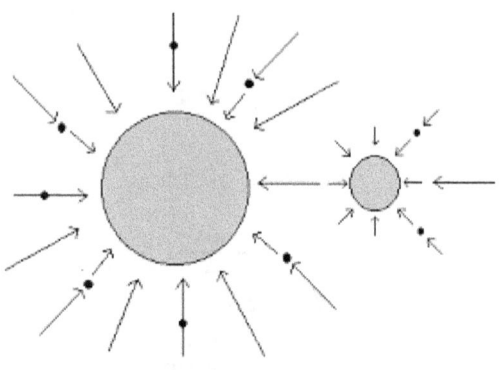

Condensing mass

The formula for the gravitational attraction of mass is:

$$F = \frac{M_1 M_2}{d^2}$$

It may be that subatomic particles, electrons, protons and neutrons are simply clouds of this condensing spinning energy in the aether. The rate of spin would be proportional to its energy and the direction of spin would be its charge. Like charges would repel each other like spinning tops bouncing off each other. Opposite charges would attract each other spinning in opposite directions moving together like gears in a massive efficient machine.

Gravity

Like Spins Repel

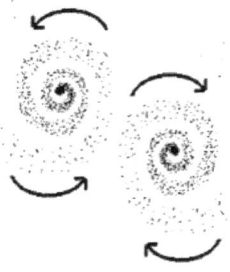

Opposite Spin Attract

Theoretically, it is these flowing and spinning clouds of aether energy, the same *medium* that transfers the waves of electromagnetic disturbances known as light, that cause gravity, electrostatic charge and magnetism. The particles of the aether do not necessarily have to flow; the "spin" may simply be energy waves trapped in orbit around these particles moving in different directions. These fields of "force at a distance" surround all mass and affect other mass in close proximity by its flow and spin without particles having to contact other mass. All of these forces decrease proportionally by the square of the distance from the mass-producing these fields. These fields easily radiate out into the open vacuum of space.

A boat crossing a river will move with the current, eventually moving at almost the same velocity of the water. The rate of acceleration depends on the boats mass and surface area to the force of the water. Unlike the force of the water on the side of the boat, the aether particles are so small that they exert a force on every particle throughout mass in general. Theoretically, every atom, electron, proton, or neutron would have a force on it from the pressures of the aether. Therefore since all mass is created of the same subatomic particles, all mass no matter what its density or size has the same gravitational force of 9.8 meters per second per second pressing on it here on earth.

Flowing River

A sailboat moves and accelerates as it is pushed by the wind and can go as fast as three times the speed of the wind; how can this be? It is not the flow of the wind that moves the sailboat; it is the pressure differential on both sides of the sail that causes movement, in fact, a sailboat if sailed properly in a zigzag fashion, can sail into the wind.

Likewise, theoretically it is not the flow of this aether that moves the object towards mass, it is the pressure differential between the higher pressures outside the central condensate particles and the lower pressures inside these central particles that cause mass to move together. The aether's differential pressure is proportional to the square of the distance to that mass, and the amount of mass. Differential pressure therefore increases proportionately closer to the attracting masses core an object comes. A large mass such as our sun will have a greater gravitational pull than our earth because the sun aether's has greater differential pressure.

Because of the earth's size, the majority of the aether flow (or pressure differential) moves through us and through all matter on the earth; every particle in every object is also absorbing this flow. Albert Michelson as we saw earlier, designed a device he thought could measure the velocity of the earth as it moved through the aether – a device designed to detect the flow of the aether wind; this was the Michelson Morley Interferometer.

As with the boat crossing the river, the current will drift the boat downstream; Michelson believed that an aether wind would light to also drift, causing his interferometer to display an interference pattern.

Interference Pattern

Michelson Morley Interferometer

As the aether wind would flow across the light beam, it would slightly displace its reflection on the 45-degree half-silvered mirror. Half of the light would go through the mirror and reflect back again of another mirror. The aether wind would then slow down the light beamed into it. Both of these light beams would then be merged again at the silvered mirror toward an eyepiece. The slight change in distance that one light beam of light would travel (due to the aether wind) would cause an interference pattern in the eyepiece, as the light waves of both beams would now be out of phase. But this was not the case and the experiment failed. Why? Because they assumed that the aether flow was parallel to the earth's surface due to the earth's motion through space. In fact, this aether flow they were looking for may be perpendicular to the earth's surface flowing into it – known as gravity.

124

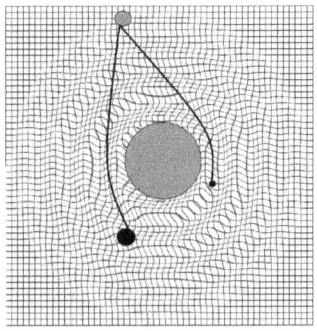

Radar Tracking of Mercury and Venus

Astronomer Irwin Shapiro's discovery while tracking Mercury and Venus's positions with radar in 1964 was in fact a modern day variation of the Michelson Morley Interferometer. Here, the beam was not light but a radar beam, and instead of reflecting off a mirror, the beam was reflected off the surface of the planets. Michelson and Morley were attempting to show that the light beam moving in the direction of the aether wind would be slowed down creating an interference pattern made by waves being out of phase. Shapiro showed that a shift in the radar frequency was created by radar energy being slowed down by the suns gravity. Here again, gravity is shown as the aether wind.

This is also evident during a solar eclipse when the corona of the sun can be seen around the shadow of the moon. This should not be the case. The size of the moon and its distance from the earth should block the suns corona totally. This was one of the first proofs that even the small gravitational field of the moon affects the movement of light around it.

125

Starlight has also been shown to bend or diffract by the suns gravitational pull, changing its perceived position in the sky. This was first seen in 1919 by astronomer Arthur Eddington while observing an eclipse off the coast of Africa (Isaacson, pg. 257).

Diffraction of Light

At present, huge interferometers such as LIGO are being used to look for gravitational waves (Bartusiak, pg. 115). What are these gravitational waves? Could these waves flow through the aether, the substance that scientists are so reluctant to admit exists?

A Star Is Born

Again, let's consider that all mass is slowly condensing material with the condensate material as the aether. As energy radiates out from the core of this cloud of aether particles, it cools and completes a phase change, turning it into an individual particle or condensate. As energy dissipates outward, the cooling material continues to condense, drawing more of this aether into it at a constant "flow". The closer to the core, the cooler this aether becomes, dropping in pressure similar to isobaric lines seen on weather maps depicting low pressure regions. The "high-pressure" regions outside the center of mass push material in towards this "low-pressure" region, spiraling in toward the center in a rotating motion.

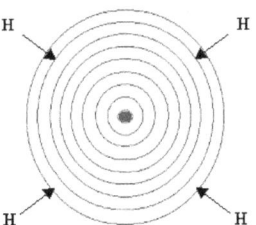

Isobaric Lines of Pressure

This process over time has caused a cooling universe to condense or coalesce as mass attracts mass and started to accumulate. If these different pressure regions did not generate during this process, the universe would stay a soup of thinly spaced particles, no light, just a dark cloud.

127

Over time, as mass attracts mass, large objects begin to accumulate more mass as their gravitational pull on smaller objects compounds dramatically. Now the large mass begins to heat up, squeezing out the energy it contains as if a sponge is being squeezed. Huge gravitational forces now cause the fusion of these fine particles into hydrogen and helium, the fuel sources for stars, and a star is born.

As the universe continues to cool, stars continue to grow until they run out of fuel, expelling all their energy. This fuel originally came from the dust and debris that the star pulled in through its strong gravitational field. Soon however, the space around the star cleared up, as if cleaned up by a gigantic vacuum cleaner and the star had to rely on the fuel it had stored within its self. Over billions and billions of years, this fuel slowly burned up, shedding its energy out into space, giving up much more energy than it absorbed. As its energy subsides, its gravitational force begins to overcome the energy supporting it from within.

At first the star grows like a balloon in a desperate struggle to survive, but soon, with very little energy left to stay inflated, it can no longer support itself against its own gravitational field; it then contracts and collapses in on its self into a black hole. The gravitational force of its huge mass has overcome the nuclear fire which once raged within it, now nothing, not even light can escape this massive gravitational field. John Michell (1724-1793), English philosopher and author of *The Philosophical Transactions of the Royal Society of London* developed this theory back in 1783 (Hawking, pg.83).

A Star Is Born

Not all stars come to this so-called end. It all depends on the mass of the star, very large stars however may indeed become black holes, and others become white dwarfs, brown dwarfs, neutron stars or pulsars. When medium sized stars like our sun collapse, they explode in a supernova, blowing off its gaseous material out into space and become neutron stars. More massive stars overcome by their own gravitational forces continue to collapse becoming black holes. Indian-born American astrophysicist and Nobel Prize winner Subrahmanyan Chandrasekhar (1910 - 1995) determined what is now known as the Chandrasekhar Limit, the threshold at where mass and energy levels allow for the development of a Black Hole (Hawking, pg. 85).

Black Hole's, once just a theory, are now popping up everywhere. It is believed now that every galaxy has one and may be what gave birth to the galaxies. As clouds of dust through gravitational forces pulled together, massive amounts of mass came together and developed mass beyond the Chandrasekhar Limit. Here mass levels were so high and gravitational forces so strong that energy levels within this congregating mass were too low to prevent it from collapsing and a black hole was born. Dust clouds in the close vicinity of the black hole started to spin at high speeds and the friction of the high velocity dust caused it to glow with enormous brightness, the creation of a quasar. The high energy and the enormous amount of mass being pulled towards the black hole generated the birth of stars, spinning at great velocities around the black hole. Dust and gasses near the center of this huge storm were slowly pulled into the black hole; the last light to escape glitters off the "event horizon", the last chance for energy to escape the massive gravitational pull of the black hole (Hawing, pg. 88).

Back to the scale of the sun, stars and planets, this continues to be the case, causing mass to attract mass; the greater the mass, the greater the attraction with a lower pressure. For example, between the sun and the earth; a higher pressure existing outside of both the earth and sun pushes them together. As light passes through this aether cloud surrounding a large mass, the density of "space-time" increases. The closer light comes to the large mass, the slower light travels, causing it to bend or refract, just as light bends while passing from air into water or glass. The stronger this gravitational field, the denser the aether (and space-time) and the more light will bend.

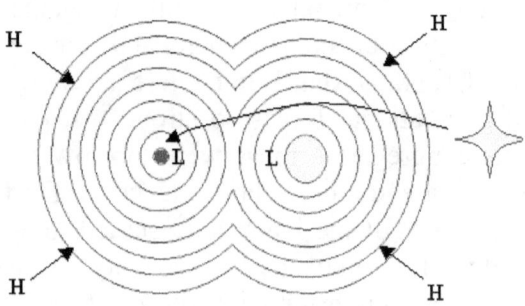

Isobaric Lines Bend Light

Black holes may actually come in many sizes (Hawking, pg. 112). The Chandrasekhar Limit, although based on the amount of mass a non-rotating object can absorb before collapsing into a black hole, also must calculate the total energy of the object. For example, when a massive star runs out of fuel, at a specific energy level it can no longer support itself and will collapse. Mass is basically constant here and energy is the variable. Smaller stars may also become black holes in the depths of space where the stars mass can drop to temperatures close to zero Kelvin. Different amounts of mass therefore will become black holes at different temperatures.

Although in today's scientific community, the concept of the aether is generally not accepted. Scientists however, do consider an invisible gravitational *field* to exist that surrounds all objects and increases in intensity with an increase in mass and decreases in intensity with distance away from the mass: the greater the gravitational intensity, the greater light is bent. What does this field of varying "pressures" consist of?

Gravity of the Matter

Magnetism

Magnetism is one manifestation of a field. An electromagnet is created with a coil of wire around a ferrous core material. When a battery is connected to the coil, an electron flow through the coil creates a field around the coils core called a magnetic field. The magnetic field contains lines of flux which seem to emanate from one end of the core, moving around to the other end, and can be seen by sprinkling iron filings on a piece of paper and putting a magnet under the paper.

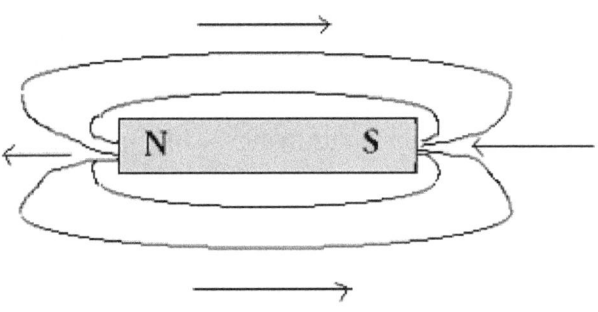

Magnet

What are these lines of force? In space or in a vacuum they exist but there supposedly is no mass associated with this flux. All magnets emanate this polarized flux or field from two poles; the north and south poles. Each pole has a charge; like charges repel and opposites attract. When a north pole of a magnet is pushed toward another magnets north pole, the second magnet will move away, repelling each other. What *medium* transfers this energy?

This field generation is similar to a radio antenna where an alternating current induces an electromagnetic field which flows with the current emanating out into space at great distances. With DC current in a wire however, current flows continuously from the negatively charged terminal of a battery towards the positive terminal; with an electromagnet, this DC current flows from the battery into one end of a coil to the other and back to the battery. As the electrons flow through the coil, the field moves with the electrons, exiting one end of the core and returning back around to the opposite end outside the coil, then entering back into the core. This is the flux seen with the filings in the form of a field. In the next figure, the flux on the outside of the coil of both the south and north poles of the magnets flows in unison drawing them together.

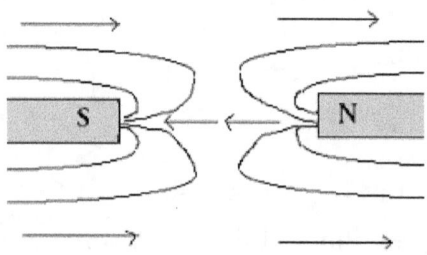

Attraction

The flow of flux out of the north pole of one magnet opposes the flow of flux out of the opposing magnets north pole. Likewise the flux flowing on the perimeter of the coil is in opposite directions. It is these opposing flux flows of the north poles that cause the magnets to repel. The same is true with two south poles of a pair of magnets. In natural magnets an electrical flow within the material itself possibly generates this field.

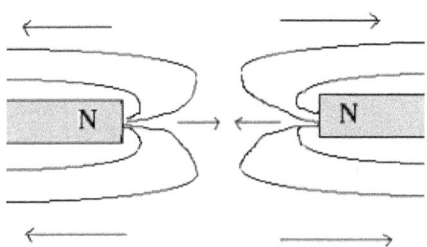

Repelling

What is this field that surrounds the magnets and works in "empty" space like gravity and also has a "spooky" force at a distance? Unlike gravity which has no polarity, magnetism like static charges has opposite forces, positive and negative or north and south.

Gravity of the Matter

Light Propagation Theory

What has been done here in this book is to show that the propagation of electromagnetic energy moves through some sort of invisible medium; could the propagation of gravity (if this is even the right term) also involve this medium or field? This supposed _medium_ that these energies move through is theorized as a gas like substance dispersed throughout space, it moves and is compressed by moving objects. Energy, as it enters into the _medium_ from an object emitting it, moves through the medium at the speed of light in reference to the movement and density of the medium. Although the velocity of the source emitting the energy may vary, the velocity of light energy through the medium is constant. The velocity of the emitter however is directly proportionate to the wavelength of the energy received at the detector, and may be perceived as Red Shift or Blue Shift.

We can imagine light propagation through this medium in much the same way we see sound move through water. Imagine two submarines in the depth of the sea sounding their sonar's in order to view obstacles. As Sub 1 is stationary in the water, it pings its surroundings with its 8,000 Hz sonar with the speed of sound in seawater at 3,355 mph producing a wavelength of 0.1875 meters. The sub begins to move and speeds up to 25 knots (28 mph) through the water. As the sonar exits its transmitter into the water, each wave moves forward at the speed of sound in water. The sub moves forward as it sends out the next wave slightly ahead of where it would be if the sub were stationary. As these waves are transmitted through the water, the subs forward velocity compresses the wavelength shorter down to say approximately 0.181 meters, closer together.

A second sub (Sub 2) arrives on the scene, and approaches Sub 1 at 25 knots. As the sound waves passing through the water enter Sub 2's sonar detector, they compress even more by the subs forward velocity through the water to say, 0.175 meters. Now, the speed of sound leaving Sub 1's transmitter was at the speed of sound to that transmitter. Once the sonar wave entered into the water, it transitioned to the speed of sound in the water and was in no way sped up through the water by the subs forward movement. As the sonar wave enters into Sub 2's detector, the sound waves transition from the water into the detector to the speed of sound in the detector. The speed of sound transitioned through three different mediums with the final speed of the sound inside the detector, in no way being influenced by the movement of Sub 1, the water or Sub 2.

Sound Energy Propagation

In fact, there could be twenty subs all moving at different speeds through the water sending out signals for the other subs to detect. Each signal, as it transitions into the water from the sub transmitting the signal, moves at the speed of sound in the water independent of the subs speed, although the frequencies may vary. Each sub receiving the signal, likewise transitions the sound into its sensors from the water. The speed of sound inside the sensor is the speed of sound in that sensor and is the same for all subs no matter what their speed through the water may be. The frequency received however will vary by the speed of that sub according to the Doppler Effect.

The speed of light also is always a component of the medium it is traveling through. As light leaves a star in the great Andromeda Spiral Galaxy, it moves toward the earth at about 50 km/s (about 30 mi/s.), transitioning from the atmosphere of the star, out into open space. The light traveling through open space, similar to sound traveling through water is in no way sped up through space by the stars forward movement. As the light enters into the earth's atmosphere, it transitions into the medium moving with the earth and on into the medium in and around the detector. The detector then detects the lights velocity in reference to the medium it is in, not the earth's movement through space. The speed of light transitioned through three different mediums with the final speed of the light in no way influenced by the movement of the star, space or the velocity of the earth.

Light Energy Propagation

The speed of light in this context is measured the same no matter what the observers and detectors velocity is on the earth, what velocity the emitter or in this case, the star is, either moving towards or away from the earth and is consistent with Einstein's Theory of Relativity. The velocity of light however, just as with sound, is dependent on the medium that it is traveling through. The perceived wavelength however, is dependent and a direct indicator of the velocity of the object the energy left and the velocity of the object detecting it. Could gravity waves also someday be detected with these same properties?

Here or There

Zeno of Elea, from the pre Socrates Greek era (late 400's BC), wrote about a race that could not be won, a paradox of sorts. In the story "Achilles and the Tortoise" a tortoise challenges Achilles to a race, as long as Achilles gave him a small head start of only ten meters. The tortoise would then argue that as Achilles would start to catch up, the tortoise would cover a little more distance of which Achilles would also have to also catch up on. This would then go on indefinitely and Achilles could never win.

We can look at this scenario another way; let's start by going from point A to point Z where each step is broken into halfway points. To get to Z from A, you must go through B the first halfway point. Once at B, you must then go through C, the next halfway point. To get to Z you must yet go through D and so on. The points keep getting smaller but you are still not at Z. Theoretically, this can go on for infinity, because you can never get to an end point unless you first go through the next halfway point. Hypothetically then, the smallest divisible points will then be Y and Z. Y being here and Z being there. There is no existence other than these positions, no halfway point. These positions could be the particles of the *medium*. When they move from here to there they move almost instantaneously, not existing unless they are here or there (Wolf Fred, pg.14).

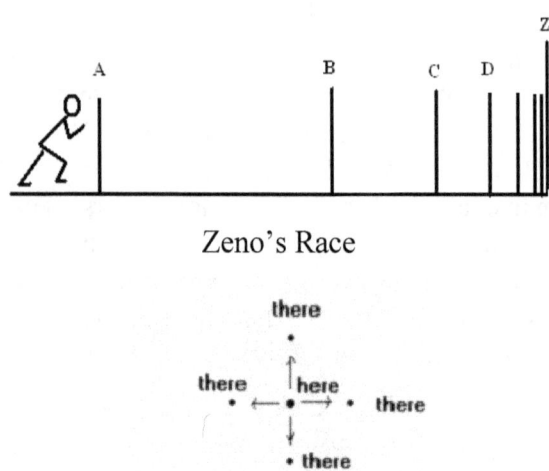

Zeno's Race

Here and There

In normal everyday life, we can measure say a car and see its location and its state of motion; this also goes for the earth around the sun and a bird in the sky. In the quantum world however, one can only know an objects position or its state of motion (Wolf, pg. 111). Because of the minuteness of this world, even the act of measurement changes the outcome of the measurement. A beam of light for example, being used to measure a particle, will cause the particle being measured to move or change. Therefore, only one measurement such as its velocity can be known, the other desired measurement must be calculated through probability. In order to mathematically calculate these motions and positions, a particle had to be portrayed as a wave function or quantum wave, according to the Copenhagen Interpretation of Quantum Mechanics (Wolf, pg. 261).

Here or There

The act of observation then collapses this wave; probabilities must therefore be used to calculate the position of a particle without directly observing it. Werner Heisenberg pictured the quantum world as a fuzzy, spread out world, a world where specifics could not be known. He stated that the wave nature of particles makes it impossible to know the particles position and momentum. In fact, the more precisely you know a particles position, the less likely you are to know its momentum, only probabilities can be calculated; this is known as the Heisenberg Uncertainty Principle (Bord, pg. 388).

Here the very act of observation changes things; even the state of the universe ultimately could be at stake with a change in observation. In fact, according to the Copenhagen Interpretation, until an observation is made, a particles position and state are mere probabilities. Therefore, nothing would exist until it is observed. Einstein would go on to say, "The Heisenberg-Bohr tranquilizing philosophy — or religion? ...Is so delicately contrived that, for the time being, it provides a gentle pillow for the true believer from which he cannot very easily be aroused" (Isaacson, pg. 453). Einstein would ridicule this philosophy as "spiritualistic" and "spooky action at a distance".

Erwin Schrödinger along with Einstein would then look for another way to describe this interaction of particles in the "quantum world". Schrödinger then devised a way to describe particles with his famous wave equation. This equation would explain the probability of a particles position and state when observed (Brown, pg. 88).

Einstein did not really like this probability "stuff". He would describe this way of thinking as taking two boxes and placing one ball in one of the boxes. Stating that the probability of the ball being in the first box as a 50% chance was an incomplete statement to him; the right way to address this was to state that the ball was in the box or the ball was not in the box; that was reality! (Isaacson, pg. 455). Many would argue that the ball was in no means in one of the boxes until they would lift the covers and observed it; the "if a tree falls in the forest" scenario. This was that "religious like" new science that many believed in and Einstein was not so fond of.

Erwin Schrödinger would come up with another analogy similar to Einstein's box. "One can even set up quite ridiculous cases. A cat is penned up in a steel chamber, along with the following diabolical device (which must be secured against direct interference by the cat): in a Geiger counter there is a tiny bit of radioactive substance, so small that perhaps in the course of one hour one of the atoms decays, but also, with equal probability, perhaps none; if it happens, the counter tube discharges and through a relay releases a hammer which shatters a small flask of hydro-cyanic acid. If one has left this entire system to itself for an hour, one would say that the cat still lives if meanwhile no atom has decayed. The first atomic decay would have poisoned it. The psi function for the entire system would express this by having in it the living and the dead cat (pardon the expression) mixed or smeared out in equal parts". It is typical of these cases that an indeterminacy originally restricted to the atomic domain becomes transformed into macroscopic indeterminacy, which can then be *resolved* by direct observation. That prevents us from so naively accepting as valid a "blurred model" for representing reality. In itself it would not embody anything unclear or contradictory. There is a difference between a shaky or out-of-focus photograph and a snapshot of clouds and fog banks" (Isaacson, pg. 456).

Einstein loved it; he would go on to say "Your cat shows that we are in complete agreement... a psi function (wave function – also shown as ψ) that contains the living as well as the dead cat just cannot be taken as a description of a real state of affairs" (Isaacson, pg. 457).

Gravity of the Matter

Time and Space

When it became obvious to scientist that the speed of light was not only constant, but also that light was always detected at the same speed no matter what the velocity of the source or detector, scientists became perplexed. Albert Einstein would go on to say that this dilemma had *"plunged the conscientiously thoughtful physicist into the greatest of intellectual difficulties"* (Einstein, pg. 18) and came up with a brilliant solution that would remedy all – warp time and space under the effects of gravity and various velocities.

There is another possibility however, that we have been subtly suggesting throughout this book, that may be more reasonable and may work just as well scientifically, another way of altering Space - Time with the existence of the aether. But didn't the Michelson Morley Interferometer disprove its existence? Doesn't research on Binary stars, gravitational Red Shift, Mercury's precession and GPS satellite clock anomalies prove that time and space warps with high velocities and strong gravitational forces?

Let's perform another thought experiment where two spacecraft are traveling at two different velocities relative to the same starlight. One spacecraft is traveling towards a star at 150,000 kps (kilometers / second) or half of light speed relative to the star. Another spacecraft is traveling parallel to the first spacecraft at 75,000 kps, one quarter light speed relative to the star. Both of these spacecraft have observers with test equipment measuring the same light but are traveling at different velocities relative to the star, yet both measure the same light at 300,000 kps.

147

STAR

photons

150,000 Kilometers / second

75,000 Kilometers / second

Relative Contraction of Space and Time

What we seem to see here is that the space and time between each spaceship and the star is being compressed and that this compression or contraction of space occurs according to the "Lorentz Contraction" of space relative to each of the spaceships velocity. According to Einstein however, **each spacecraft** would have a different rate of time passage and different measurements on board due to their individual velocities relative to the star; it is not space between the star and the spaceship that is altered, it is the ship itself.

Now, if one spaceship would come between the star and the other spaceship blocking the starlight, suddenly the focus is different. The light the trailing space ship then is viewing is the light reflected off the spaceship in front of them traveling at a slower speed, rather than the light of the star. Did the space ship suddenly shift its shape to adjust to the light reflected off the ship in front of it? If this were true, the ship would have to reshape according to all objects giving off light and moving at different velocities relative to it, compressing for some and expanding for others. Similarly, for binary stars moving in opposite directions to an observer, would both stars stretch and compress at the same time to accommodate for the speed of light to different observers in different regions of space viewing them? Of course not!

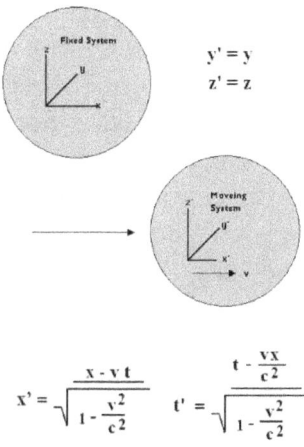

$$y' = y$$
$$z' = z$$

$$x' = \frac{x - vt}{\sqrt{1 - \frac{v^2}{c^2}}} \quad t' = \frac{t - \frac{vx}{c^2}}{\sqrt{1 - \frac{v^2}{c^2}}}$$

Lorentz Contraction

The velocity of the light detected from an object can only then be relative to the observers themselves; the observer's actions cannot affect the motions or observations of other observers and the space-time around them. Only viewing velocity, time and space in this matter prevents a paradox. This also allows for any velocity to occur without changing space and time; therefore an observer's time and space cannot change with their velocity as previously thought. If time and space is not being warped by the movement of an object, how then is the speed of light the same velocity to all observers, no matter what their velocity or the emitter's velocity?

This brings us back to the submarine illustration. The received sonar signals speed inside the sonar dome of the receiver is at the speed of sound and is constant inside that dome, no matter what the speed of the sub emitting the original signal, or the speed of the sub receiving the signal. The velocity of the water currents between the two subs also has no effect on the sounds velocity inside the sonar dome after it is received. Likewise, the velocity of sound coming down the ear canal of an observer listening to the trains whistle also is unaffected by the trains velocity or the observers own velocity. Sound always has the same velocity; it is only changed by the temperature and pressure of the air it is traveling through. The only thing that can change due to the movement of objects then is the frequency of the detected sound, the Doppler Effect as previously reviewed.

The same must be true with light. The velocity of light inside the detector, whether it was Bradley's telescope, Fizeau's and Foucaults instruments, or the Michelson Morley Interferometer will always be the same. This velocity is always 186,000 mps or 300,000 kps, no matter what the speed of the emitter (train), movement of space between the emitter and receiver (wind) or the movement of the receiver (observer). This light moves in waves within some rigid medium and will only vary in velocity depending on the mediums density; only the received lights frequency will vary according to the movement of the various elements as seen with the Doppler Effect. The two spaceship previously discussed will go through the same process. It is not the spaceship that contracts or the space between the star and the ship that contracts, it is the light waves themselves that change as they move from one medium to another. The speed of light inside the detector remains the same, no matter the movement of the ship or the star – just like the submarine.

Time and space here are still affected. The speed of light will vary according to the type of medium it travels through, for example light traveling through glass as compared to light traveling through air. The time it will take light to travel a specific distance will then vary; space also will then vary proportionately with density. Space then also becomes denser close to mass through the cooling process or phase shift as previously discussed and less dense the further away from mass.

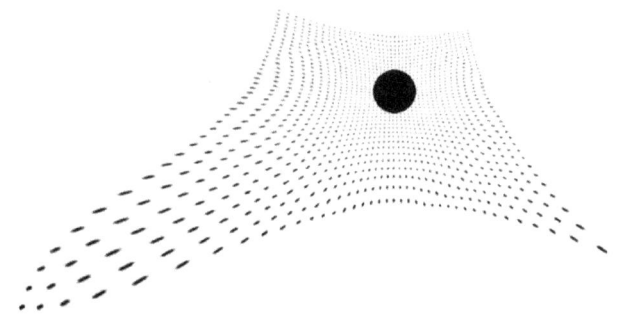

Warped Space

Einstein made an analogy using a bowling ball on a trampoline to describe how space warped with large masses such as our sun. The bowling ball would cause the rubber sheet of the trampoline to sink in the middle, as objects

151

such as billiard balls were rolled onto the sheet; they would roll inward toward the bowling ball (Isaacson, pg. 223). This analogy was Euclidean or two dimensional in description, in real Space-Time there would be an infinite number of sheets surrounding this ball in three-dimensional space with many layers. In three-dimensional space, it would be spherical sheets much like that of the layers of an onion. Each of the layers of onion would represent a change in gravitational field strength, a change in field strength density if you will. The closer to the bowling ball, the tighter together the sheets become and likewise, the further away the more distance between the rubber sheets or onion layers.

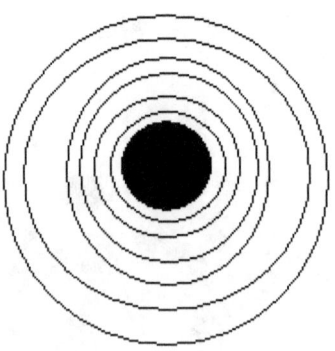

Onion Analogy

What is this field made up of that changes strength? What would change in density? As we reviewed waves in previous chapters we also asked the question, "What was waving?" Could it be the same material or medium theorized as the aether?

Time and Space

In the previous chapter "Here and There," we reviewed that if we were to keep dividing distances in half, which eventually we would come to a point where division was no longer possible; a particle was here and another particle was there, nothing was in between. We can look at these particles as the field that exists in "empty" space, the medium that transfers energy across great distances, and whose condensation creates elementary particles. Could it be the variations in the density of these particles that cause light to bend and the differences in energy levels that create the pressure differential that causes gravity? The pressure differential within each of these gradients would be detected instantaneously therefore preventing the "gravitational aberration" as discussed earlier and curving Space-Time. This could be the onion we are looking for?

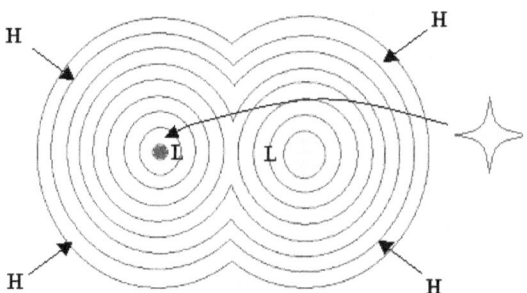

Isobaric Pressure Gradients

An example of this is the speed of sound in water. At 100° Celsius (boiling), the speed of sound in water is 1,543 meters per second; however at 0° Celsius, (freezing) the speed of sound is 1,402 meters per second. The closer together or the denser these particles are, the longer it takes for energy to pass through them; this is the principle of refraction of light in thermoclines (sharp changes in temperature) in water and also possibly is what causes light to bend around large objects in space. The closer to the earth, sun or any other condensing mass such as subatomic particles, the denser the medium becomes that light waves move through; the denser the medium the slower the energy moves which also effects "local time".

Warped Space-Time

If mass is merely created out of standing waves within this medium, then the velocity of these waves will also slow down as the medium becomes denser. Motion will be dampened, clocks will tick slower and redshift will be noticeable in light waves emanating out of these objects with large gravitational fields - in fact time will slow down inside of these gravitational fields. This gravitational effect has been clearly noticed in the timing circuits of the GPS satellite system.

The larger the mass, the slower time moves close to it; likewise the deeper into these onion layers an object moves, the slower the time. Super massive objects such as black holes may indeed slow time down so much in the center that time virtually stops and a singularity similar to the one that existed at the beginning of the "big bang" could exist. The difference here is that this is a miniature "big crunch" or a collapsing of matter, rather the massive expansion of matter during the "big bang." If time goes slower within the stronger gravitational fields, then the reverse should be true outside the gravitational effects of large masses – time should then speed up theoretically to infinity (Hawking, pg. 90).

The "Theory of Relativity" also states that velocity also changes time. Previously we reviewed relativity and how that a moving object's velocity is only relative to the position of another object. For example, if a spaceship was far out in the universe with no object to truly reference velocity by, and came across two objects such as asteroids moving away from each other, what would be used to reference the speed of these objects? If they were floating apart at 500 meters per second, was object "A" moving at that speed away from a stationary object "B"? Or was "B" moving away from stationary "A"?

Space Objects in Motion

If "A" was moving faster than stationary "B", time would be slower on "A" but if "A" was stationary and "B" was moving away then "B's" time would be slower. Velocity here is relative and therefore velocity can have no effect on time on the moving objects. There is one catch however, if both objects were stationary relative to each other at one time and one began to move, the object that started to move away had to accelerate. The force of acceleration in the "Equivalence Principle" is the same as gravity – they are indistinguishable. Therefore, it is the acceleration of a moving object that must give it its time change, not velocity itself; once an object stops accelerating, this effect disappears.

Unless a single point of reference for all objects is determined, velocity is simply relative. Only when we have determined that this elastic solid medium that energy waves are moving through exists, then do we have a reference to measure the speed of light to – **a stationary point in space**.

Velocity, no, rather a change in velocity, well, let's say a constant change in velocity due to acceleration will then slow down time. The longer and the stronger the acceleration forces, the slower time becomes. Therefore, if an astronaut launches out into space and accelerates up to great velocities, time on earth will pass faster than time on the spacecraft. When he arrives back, more time will have passed on earth than in the space ship – he would be younger than if he had never left. If an astronaut has to accelerate leaving the earth, he will likewise also have to decelerate to slow down and return. This deceleration as we saw before, simply changes the direction of the force from aft to forward. The effects of deceleration are identical to acceleration – they therefore are identical. As long as this force exists, time will slow down. Why leave the earth then, why not use a spinning acceleration machine instead to change time? The amount of high forces due to acceleration or gravity may change time but it may be so negligible that it would never be noticed.

We have seen that time may be slowed down due to the effects of gravity or acceleration. Can time be reversed though? Can we go back in time? In his book, A Brief History of Time, Stephen Hawking suggests that even though the Theory of Relativity states that speeds faster than light are impossible, if one could do so, he probably could return before he left (Hawking, pg.163). Simple math however prevents this. Much like our analogy of Zeno's Race, no matter how much you multiply a velocity by, you can never get to zero time. For Example, at 100 miles an hour, it will take you one hour to reach the one hundred mile mark. You multiply your velocity by two; you will reach the one hundred mile mark in a half an hour. Multiply by three, one third of an hour and so on. No matter how fast you go, you cannot go faster than time.

Time is not the only thing that changes; space also is changing (which in effect changes time). We studied the possibility of the "Big Bang" how that the universe and all that we know originated out of a singularity, a point where time and space were "infinite". A huge blast occurred when the natural balances that contained the singularity broke down, blasting matter and energy out into the far reaches of the universe. Energy, not being able to exist without mass (it is simply the movement of mass) must be associated with some sort of particle, the particle that all mass consists of, the "God Particle". These particles originally were spread out evenly throughout the universe, but as energy was passed on from one particle to another, this energy was dispersed out and away from the center of the blast. Higher energies to the outside of the blast area forced cooler particles in the center of this area to coalesce and bond together, a phase transition of some sort. This was the foundation blocks of the stars and galaxies that we see today.

As energy slowly is dispersed out and away from the original blast sight, the center continues to cool and contract. Stars and galaxies further away from the center of this blast should therefore have more energy, expanding faster than the cooler areas. Over time, the whole universe should be also cooling uniformly (although areas further away are hotter). Areas further away however, also take longer to view from earth so that objects further out are seen at higher temperatures and at velocities than what they may be today; these areas have since cooled down and slowed down, possibly dramatically.

Edwin Hubble's discovery in 1929 suggested that galactic Red Shift in light from stars and galaxies increase in proportion to their distance from earth. His studies showed that 38 of 40 galaxies observed are rapidly moving away from each other, an omni-directional galactic expansion where galaxies furthest away are moving the fastest, some at 57,000 kilometers per second or 13 million miles per hour or 2% the speed of light. The problem here is that this is evidently **omni-directional**, indicating that we are the **center of the universe**. If we were viewing this expansion out on the edge of the universe, only certain areas focused in one region would have the greatest Red Shift. Either we are the center of the universe, or this observed redshift is simply an **illusion**.

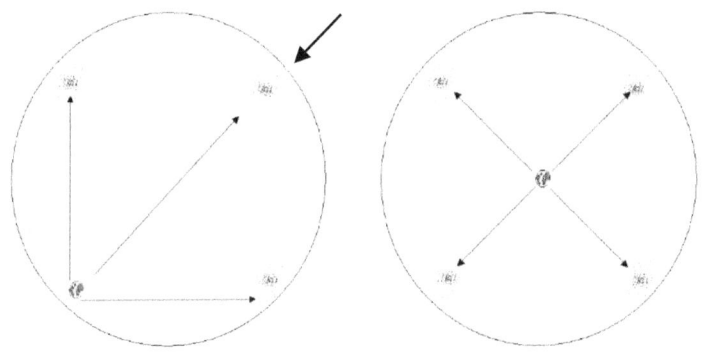

Center of the Universe?

Since it takes time for the universe to cool and also takes time for light to reach us on earth, the changes to the universe will be isotropic or seen uniformly in all directions and be relatively consistent at the same distances because of time. This however suggests a possible **"Big Crunch"** as energies subside and particles coalesce. Eventually as the medium between all objects condenses and cools, there will be nothing to keep objects apart. Higher energies outside will eventually push all matter together into one place. Here, at this point, atoms can no longer exist, fusing together into what has been called the Bose–Einstein Condensate or BEC's, huge molecules that form at near 0 Kelvin or absolute *zero* – cold black holes if you will (Pitaeskii, pg. 1).

Energy however, would have to go somewhere; it cannot just disappear according to the Principle of the Conservation of Energy. If space is infinite in all directions, over time all energy would be evenly dispersed omni directionally and the universe would eventually grow cold and die. Now, if we take the stand that all other matter, including ourselves, were merely energy waves (standing waves) moving through a stationary substance that varied in density and energy, our existence would be limited. The further out in the universe we would go, from the center of the Big Bang, the less dense this substance would be, to the point where matter no longer could exist; much like how sound cannot propagate and exist out in space. Energy cannot move through nothing, therefore energy would have to slowly migrate back into the known universe, reflecting back in from the edge of Space-Time if you will, pressing in on mass creating gravity.

So, if energy was forced to loop back in at the edge of the universe or loop back around though black holes and singularities into other universes that constantly feed each other, then it is possible that matter and energy may be able to continue on indefinitely. Many different theories have been developed over the last fifty years trying to show how this all may work, from a universe in the shape of a doughnut where time and space loops back in on itself, to the "String Theory" of multiple universes. If we think about it, a black hole's singularity draws in matter; the singularity that occurred at the "big bang" gave off matter. Can one singularity be the source of the other singularities energy? Is this the entrance to a parallel universe?

Gravity of the Matter

Cosmology

Orbital velocities of bodies in our solar system, normally vary according to Keplerian dynamics; orbital velocities that are proportionate to the distance from the object they orbit. The further away from that object, the slower the orbital velocity of the planet or moon, decreasing this velocity inversely with the square root of the orbital radius; this is due to gravitational forces decreasing by the square root of the distance. Mercury for example, takes 88 earth days to orbit the sun while the earth takes 365 days. Mercury's orbital velocity is 48 km/s while the earth's orbital velocity is slower, about 30 km/s in it voyage around the sun. Likewise, Neptune's orbital velocity is only 5.4 km/s and it takes about 164 earth years to orbit the sun.

This, however, is not the case with stars orbiting in a galaxy. Surprisingly, most stars orbit a galaxy in unison, as if tied together; stars on the outer fringes are therefore are orbiting much faster than they should, some at a radial velocity as high as 180 km/s and theoretically should be slung off into space. As with interactions within our own solar system, orbiting objects are not only attracted to the object they are orbiting; they also have gravitational forces acting on them from all other objects relatively close to them. The earth is not only attracted to the sun, but all the other planets including our own moon, causing the orbital distance from the sun to vary. Stars likewise, are not only attracted to and orbit around a black hole in the center of the galaxy, they also are gravitationally attracted to all the other stars, planets and gasses moving around the black hole, as if in a soup, moving together. Calculations indicate that there must also be a missing mass that can't be seen in order to account for the necessary increase in gravitational forces needed to prevent these stars from flying off, this is the foundation for the **Dark Matter** theory.

Further calculations had to be formulated to account for additional missing matter, as only 25% of the universe's matter could be accounted for. This was the foundation for the **Dark Energy** theory; the energy that helps accelerate stars and galaxies in the far reaches of the universe, faster away from each other, as manifested in Hubble's Law. This law states that the velocity at which galaxies are moving away from the Earth is proportional to their distance from the earth. The cosmological formula most recently calculated for the universe's total mass is 4 percent baryonic matter (composed of atoms and subatomic particles of some sort), 21 percent Dark Matter, and 75 percent Dark Energy. How does this work then? What are Dark Matter and Dark Energy? Let's review the Big Bang.

As discussed previously, the Big Bang theoretically occurred some 10 to 20 billion years ago when something happened within a singularity and thermonuclear energy within the singularity could no longer be contained. A huge amount of energy was released and the smallest of particles blasted out into space at enormous velocities, dissipating their energies out into the blackness of the universe. As this outward energy dissipated, a reversal of the flow or an "implosion" of these particles of mass begins to take place, back toward the center of the blast and towards each other, slowly condensing these smallest of particles into baryonic matter. The outer regions of the universe now contain much of the energy released in this blast as the origin, or center of the blasts begins to cool. At some point in time, the energy blasted out into the universe greatly diminishes and dissipates as the dispersion of energy decreases by the square of the distance it travels away from its source.

The question now being asked is, 'are particles and objects closer to the Big Bang's source moving back towards the origin of the blast, as if sinking into a Black Hole?' If this is the case, could these objects (galaxies, stars, and even us) be "moving away" from the galaxies at the fringes of our universe, rather than those galaxies "moving away" from us? Theoretically, the further away from the origin of the blast, the weaker the pull is on objects towards the center of the Big Bang, giving the illusion of variance in velocity in relationship to distance. As matter condenses and becomes more compact, time also slows down, giving the illusion that galaxies seen at the fringes of our universe (looking back into time) are moving away much faster than they currently are in reality. Could it be that the Universe is no longer expanding and that Hubble's Law is an illusion; an illusion created by an accelerating collapse of "Space-Time"? The Big Crunch may be underway!

If Hubble's Law were simply an illusion, Dark Energy, the substance that supposedly accelerates galaxies at the edge of our universe, would no longer be required. All that would be needed would be Dark Matter and baryonic matter, materials consisting of atoms and subatomic particles. Dark Matter possibly could be the substance that all baryonic matter is made of, also known as the aether, the medium in which all energy moves through.

Hubble's Law – The further away objects are from earth, the faster objects (galaxies) move away from the earth, causing Red Shift. The earth would then theoretically have to be located at, or close to, the center of the Universe where the Big Bang occurred.

Inverse of Hubble's Law – For closer mass is to center of the Universe where the Big Bang occurred, the faster mass moves towards the center (as if falling into a Black Hole). For us on earth this creates Red Shift as well as the illusion that further away objects are moving faster away from us.

Cosmology

Triangulation has been used for centuries to calculate distances or the height of terrestrial objects, such as trees or mountains. A known length, or height, at a known distance, is compared with an unknown dimension at a greater distance in order to calculate an approximate measurement; a pencil at arm's length compared to a tree for example. One observable variance similar to triangulation with stellar objects is called parallax. Parallax occurs when our viewpoint relative to nearby and distant objects changes, creating an apparent angular change in position of the nearby objects relative the distant objects. This effect diminishes with distance so that even for the stars nearest to us, the parallax is so small that extremely precise measurements are needed just to detect the effect. Due to the earth's rotation around the sun, the distances to planets such as Jupiter (although it also moves in orbit around the sun) can be approximately calculated. Parallax of some stars can be observed as the earth moves in orbit around the sun over time; here stars closest to us in our own Milky Way galaxy seem to wander slightly, when compared to the stars around them that are further away.

This process however cannot be used to calculate the distance of distant stars and galaxies. Astronomers must use luminosity to calculate the distance to an object, such as a galaxy, by measuring the brightness of the object and applying the inverse square law. The problem is stars vary in size, color and brightness; even galaxies vary in these attributes and are almost impossible to calculate unless you find a standard brightness or candle to go by.

The best standard candle for determining the distance to the nearby galaxies is the Cepheid variable star; a star that has a precise level of luminosity and is very common throughout the universe. Some of these stars are even close enough, in our own galaxy, to measure their distance by parallax. With a known distance and a known luminosity, the distances to other Cepheid variable stars in other galaxies can now be calculated. Edwin Hubble used Cepheid variable stars that he recognized in the Andromeda galaxy (our closest neighbor), to approximately calculate its distance from us at 2.5 million light-years. While cataloging the distance to these various galaxies, Hubble discovered that the light signatures from the furthest galaxies away from us had the most Red Shift; varying proportionately with their distance from us, thereby establishing Hubble's Law.

String Theory

String Theory is one of the latest attempts through very detailed mathematical calculations to explain quantum mechanics using vibrations of one-dimensional strings, which only have length. These strings are the constitutes of all particles and are the foundations of time and space. Various vibrations at different harmonics, such as the strings on a violin that vary in pitch and amplitude, can represent various components of subatomic particles such as quarks. Different combinations of these strings vibrating at different frequencies manifest themselves as electrons, protons and neutrons. Combinations of these strings are called membranes or "**branes**" for short. Different branes may represent different universes and the collision of these branes, like sheets hanging on a clothesline bumping into each other, may be the source of the "Big Bang". Strings can be open or closed; one closed string manifests itself as a graviton, the source of all gravity (Hawking, pg.176).

Open and Closed Strings

Similar to Theodor Kalusa's five dimensional theory proposed in 1919 (Isaacson, pg. 337), M-Theory envisioned in the late 1990's, is one of the latest string type theories that developed 11 different dimensions, 10 spatial and one time: M standing for membrane. Super String Theory is the latest attempt to combine and synchronize the various string theories (five seemingly different theories) into one. It now seems that all of these theories were all explaining the same thing in different ways, all being supposedly correct. All of this was developed with huge mathematical formulas based on the existence of a one-dimensional and two-dimensional world. Equations in the three and four-dimensional worlds became much more complex, explaining the eleven dimensions even more so (Ashtekar, pg. 316).

There is a very strong similarity between what has been suggested in this book and the **String Theory**. We have looked at the possibility that all that exists simply exist as energy waves manifested as **standing waves** moving through some semisolid substance. Solid because as far as we know, transverse waves of any kind can only pass through solids and cannot pass through fluids; light waves are clearly transmitted as transverse waves having polarity. Instead of minute particles oscillating at various frequencies and amplitudes as energy moves through this substance, the String Theory suggests that small one dimensional strings of various lengths both open and closed in structure, vibrate at various frequencies and amplitudes to create the same effect –our world as we know it – a coincidence?

170

Unified Theory

Until recently, scientific laws have basically obeyed Newton's laws. Scientific discoveries of the 20th century however have uncovered a new science, one that does not seem to obey the same laws. As these different theories evolved, we have slowly developed one sort of law structure for the large and another for the very small. These two theories are somewhat different, as Tim Allen in his book, "I'm Not Really Here" states that the rules, "at a subatomic or quantum level are, well.... different" (Allen, pg. 10).

In his later years, Einstein struggled to make the quantum world and relativity compatible, a struggle to mathematically unify the gravitational and electromagnetic field theories into on **Unified Theory** (Isaacson, pg. 337). "It has been my greatest ambition to resolve the duality of natural laws into unity", Einstein said, "The purpose of my work is to further this simplification and particularly to reduce to one formula the explanation of the field of gravity and of the field of electromagnetism" (Isaacson, pg. 342).

Einstein struggle to his deathbed trying to find the answer for the Unified Theory, writing paper after paper, trying to solve the riddle; but page after page seemed to be flawed, losing the "physical" explanations of his older writings, moving towards "pure mathematics." Scientists of that day began to lose their passion and trust in his writings, and Einstein began to lose his touch, being rejected by many of his colleagues and became skeptical about his work.

171

Various forces are apparent in the universe, as we have discovered. These are electric fields, magnetic fields, electromagnetic fields, the strong and weak nuclear forces and gravity. All of these forces seem to hold both positive and negative characteristics; pushing or pulling, except for gravity. Combining these two theories into one became a nightmare! They just didn't work together.

Classical mechanics and Einstein's Theory of Relativity, although there are some things that seem to conflict, work together much better than the theories of classical mechanics and quantum mechanics; they just did not work together! The **Heisenberg Uncertainty Principle** for example would not work well with classical mechanics. This is where the more precisely you know a particles position, the less likely you are to know its momentum, only probabilities must be calculated.

As we look back at the ideas generated in this book, we see that even through the simplicity of classical mechanics, gravity can be created. Going back to the "Big Bang" and the enormous energies that were released, to the cooling processes that caused mass to coalesce from the building blocks of nature (possibly strings), and the high energies outside of mass that draws particles together.

We also saw how as objects and particles spin, that particles within "empty" space between these particles also may spin (or at least the energy moving within these particles) affecting other objects at a distance. Opposite spins allow objects to move together and not repel while objects of like spins repel each other.

Unified Theory

Building this type of universe where laws work together and that both small and large, the regular world and the quantum world move together is not out of the question.

Gravity of the Matter

Worm Holes

We have developed a hypothesis that a medium exists that energy waves move through; a medium that varies in density according to its "temperature" or "pressure". As this medium cools, it condenses, forming particles that are the foundation blocks of all matter; the closer to matter the medium is, the cooler and denser this medium becomes, in effect changing time and space. We have also hypothesized that matter may also be created by "standing waves" of energy passing through this "elastic-solid" medium. Close to the condensing mass, these energy waves move slower and tend to bend as they pass large bodies of mass, refracting if you will, distorting space and time. The closer to the large mass these standing waves come, the slower they move. This difference in density, creates a pressure differential which in turn causes the gravitational pull; the greater the mass, the greater this pressure differential and gravitational force.

The opposite may be true as these waves move out and away from large masses such as stars and galaxies. Just as light velocity increases as it moves out of the medium of a piece of glass in to air for example, energy moving through "near space" at the velocity of "c" may move much faster far away from large bodies of mass. The velocity of light in space has been assumed to be the constant "c", 186,000 miles per second. However, if space is not empty as we have believed and light waves move through a medium that changes in density, the speed of light will also change according to its density of this medium it passes through.

If we picture space as a large ocean full of islands, we may also envision an ocean full of currents. As ships on the high seas use these currents today, we may likewise also be able to use the currents in outer space. Areas of space between stars or between galaxies may contain regions where space is almost void of this medium, allowing energy or mass in the form of energy (standing waves) to pass from one region to another at very high speeds – in effect, passing through a natural worm hole.

These wormholes may also possibly be generated mechanically. High beams of energy may be able to temporarily create a void in space, similar to how a lightning bolt superheats and ionizes air momentarily, followed by the sound of thunder. The momentary opening of this tube, mostly void of the medium would allow energy to move through it at speeds much higher than the accepted speed of light of "c", allowing us to "jump" from one region of the universe to another.

This "jump" may also be used for short distances, allowing for transportation from one area on earth to another instantly, "beaming" people or objects from place to place. Energizing or "heating" the medium in close vicinity to an object would decrease the mediums "pressure" on one side of the object. This pressure differential would cause a "gravitational pull" if you will, in a specific direction, moving the object without G forces, as if "falling" towards the targeted area. If the medium were energized below an object, it would cause the object to rise and be weightless. An object used as a transportation devise could be moved in any direction, and at almost any speed, without the adverse effects of G forces and acceleration /deceleration. How to energize the medium or aether in order to cause this "G-less" acceleration or "falling" at almost instantaneous speeds is the question. Specific wavelengths of electromagnetic energy may be the answer, similar to how water molecules are energized in a microwave oven. Could the frequency of Cosmic Background Radiation be a key to creating a Wormhole?

Gravity of the Matter

Black Holes

The current Webster's Dictionary definition of a Black Hole is *"a celestial object that has a gravitational field so strong that light cannot escape it and believed to be created due to the collapse of a very massive star"*. The scientific theory of a Black Hole is relatively new. As seen previously, geologist John Michell wrote about the idea over two hundred years ago in his 1783 paper, "Philosophical Transactions of the Royal Society of London". Here, Michell suggested that, if a star were to grow to sufficient mass, at some point, its gravitational force would be great enough to produce so strong of a gravitational field that even light could not escape from it. It wasn't until 1969 that the term was coined when scientist John Wheeler used it in a speech, and stuck (Hawking, pg. 84).

Soon, new ideas began to surface and Black Holes started to become more scientifically complex. The core of the collapsed star now had a "singularity" similar origin of the Big Bang and the outer shell of the Black Hole was now called the Event Horizon. New studies of galaxies showed high velocity stars spinning near the center of these galaxies, and x-ray emitting gasses traveling close to the speed of light, seemingly proving that Black Hole's existence (Bartusiak, pg. 210).

In his book, 'Hawking on the Big Bang and Black Holes', Stephen Hawking explains empty space as not truly empty, but full of virtual particle pairs consisting of both positive and negative energy. Hawking goes on to explain that as these particles begin to be pulled into the Black Hole's Event Horizon, the pairs split up; the positive particle being ejected out from the Black Hole as "thermal emission" while the negative particle "tunnels" uneventfully into the Black Hole (Hawking, Hawking on the Big Bang, pg. 88).

Although Hawking does not subscribe to the "Aether Theory", he seems to indicate that "empty space" is not empty after all and that "local energy density" varies with varying amounts of these particles from place to place (pg.87). This strongly parallels the hypothesis that is being promoted by this book that

(a) All matter consists of condensates of some material
(b) Energy cannot exist on its own and is simply manifested by the movement of this material

Hawking also suggests that the positive energy particle is being ejected away from Black Hole as "thermal emission" or energy radiation of some sort. Could it be that Black Holes are simply made up of the predicted, Bose-Einstein material, which is created by mass so super cooled that it collapses into its self as one solid material? Super cooled due to the fact that matter has all the energy "wrung out of it" as radiant energy, before it falls into the Black Hole?

Theoretically, as mass approaches a Black Hole, energy literally is squeezed out of it; evidenced by the x-rays and the brightness of the center of galaxies such as MCG6-30-15. Here, it would not be the attraction of the super massive Black Hole that pulls in all material via the huge differential in pressure between energetic particles in the universe and the cold condensed particles in the Black Hole. It is the universe pushing in that creates gravity, not the mass of the Black Hole that pulls in on the matter surrounding it. As energy is squeezed out of these particles, the particles move slower and slower until all energy subsides and unifies itself with the huge Bose-Einstein mass of the Black Hole.

The Black Hole is therefore "black" not because of its intense gravity that prevents light from escaping, but simply because it has no energy at all; as all energy is forced out of mass before it enters the Black Hole. Again, this suggests that the universe is cooling and contracting as energy bleeds away from its core. The expanding universe may again, simply be an illusion, and the Big Crunch is under way.

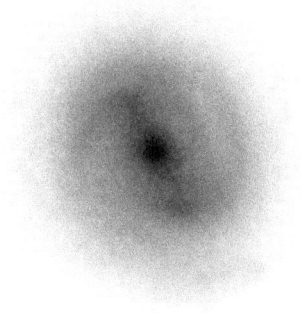

Gravity of the Matter

Review of Literature

Where did we come from and what are we made of? Man has been curious about what light is, its composition and how it worked has always intrigued us, even back to the 300 BCE's, in the days of Plato, Aristotle and Socrates. In those days, light was considered the fifth element; the "luminiferous aether", the stuff stars were made of. It wasn't until Galileo in the early 1600's that the questions of lights speed was tested with the use of covered lanterns. All that he could determine at that time was that the speed of light was at least 10 times faster than the speed of sound. Not too much later in the late 1600's however, Ole Rømer determined through the study of Jupiter's moons that the speed of light was be 299,792 km/sec or 186,290 miles/sec, becoming the cornerstone of quantum physics, the study of atomic matter.

Various theories of what light consisted of, slowly developed, including René Descartes undulation theory, otherwise known as wave theory in the early 1600's and Sir Isaac Newton's corpuscular theory or particle theory in the early 1700's. Newton's particle theory however did not stand very long as Thomas Young in the late 1700's used the "Double Slit" experiment where light created interference patterns, proving that light was composed of waves. Studies showed that light was made of both longitudinal waves and transverse waves complicating the aether theory of the day. This suggested that the medium that light traveled through, must be created of a solid substance that would resist the motions of the planets and the stars and did not seem to make sense.

In the 1700's, James Bradley became the first to prove the Copernican heliocentric view of the universe. It became evident through his studies when viewing stars through his telescope that the earth moved very rapidly though space in its orbit around the sun. These studies of star aberration, or the apparent shift in a stars position, showed different displacements of the stars over time in the northern region around the North Star and south in the equatorial region. This lead to his calculation of the speed of light which was very close to Ole Rømer's calculations 100 years earlier. This knowledge was reinforced by Fizeau's and Foucault's experiments showing lights speed within a couple thousand miles per second.

Later on in the late 1700's, François Arago tried to show that light traveled at different speeds due to gravitational forces of stars and the movement of the stars. His experiment, using a prism and a telescope was set up to detect these various speeds using Newton's refraction theory. The test failed; at least in showing that there was no variation in the refraction of light. This experiment added confusion to the mix by indicating that the speed of light from large stars and small stars, and that of moving stars in relation to the movement of the earth was the all same.

The medium, through which light moved, at this time, was thought to be dragged along with the earth through space. Star aberration however disproved this, showing that the aberration was due to lights displacement by the earth's movement and was not dragged along by the atmosphere. French physicist Augustin Fresnel would then develop the partial aether drag theory, suggesting that light traveling through air would not be dragged along, but that as the refraction index of a substance increased, so also would the drag effect increase. Light would therefore travel unimpeded through the air and be uninfluenced by the movement of the earth. Once light enters into glass or some other medium of higher refractive indexes it would therefore be constrained or locked into the new medium and travel at one velocity with that medium.

George Biddell Airy later in the 1800's, tried to detect the aether using a kind of water filled telescope. The theory was that light traveled slower through water and that while observing star aberration due to earth's movement through space, the aberration angle would change. This also failed; indicating that the aether or medium though which light traveled did not exist adding perplexity to this study. In the late 1800's Albert Michelson and Edward Morley went on to disprove the existence of the aether with their interferometer experiment.

Willem de Sitter, early in the 1900's discovered that light traveled the same speed in his studies of twin stars; much like Arago's experiments in the 1700's. His discoveries showed the Doppler Effect evident in the shifting of the light spectrum towards the red or blue due to the movement of these stars. Although the speed of light remained the same to the observer, the frequency of the light viewed changed proportional to the speed and direction the star was moving in reference to the observer.

185

These experiments seemed to confirm that there was no aether; not only did these tests show that there was no aether, they had also discovered the "Law of the Constancy of the Speed of Light." This measurement of the speed of light is the same no matter what the observer's velocity is, what velocity the emitter is, either moving towards or away from each other expressed as "c" and is measured at 299,853 km/sec. or 186,328 miles/sec.

Hedrick Lorentz would then try to rectify this strange problem by developing the "Lorentz Contraction" theory. This theory showed that the movement of an object causes length contraction of that object in the direction of its movement, therefore causing the speed of light to remain constant no matter what the movement of the observer or object being observed. This is one explanation in how the Muon can reach the earth at high speeds even though its life cycle supposedly would prevent it. The "Lorentz Contraction" due to its velocity at near light speed, allowed it to reach the ground.

Albert Einstein would then build on this to develop his "Theory of Relativity". Einstein theorized that movement changed time and space and that the faster an object moves, the more time and space around it would change. Gravity also would have a dramatic effect on space and time causing light waves to bend around stars and for time to vary in different gravitational fields. This explains the constancy of the speed of light and the shift in the light spectrum due to gravitational fields and light from objects at different velocities.

186

Modern day studies seem to verify Einstein's theories beginning with studies on the precession of Mercury. Studies of starlight being bent by the suns gravity, gravitational lensing due to the intense gravitational fields of galaxies also validates this theory. Galactic redshift and gravitational redshift is seen in the GPS transmission or radio waves and even the tracking of Venus and Mercury with radar. It seems that no matter how you look at it, everything seems to indicate that Einstein was correct.

Edwin Hubble's discovery in 1929 showed that while studying galactic Red Shift that redshift in light from stars and galaxies increase in proportion to their distance from earth. This developed into what is now known as Hubble's Law, expressed by the equation $v = H_0D$. The further we look out into the universe, the faster galaxies and stars accelerate away from each other, indicating that these galaxies and stars were moving out from a single point in space over a long period of time.

This discovery lead to the "Big Bang Theory" as scientist began to work back in time to a point where this rapidly expanding universe theoretically began, to a one point in space and time called the "singularity" over 13 billion years ago. In 1964, Arno Penzias and Robert Wilson would discover Cosmic Microwave Background Radiation (CMBR), evidently a relic of the "Big Bang" dispersed evenly across the sky.

Studies from observing these stars and galaxies generated new ideas, one being Dark Energy. What was the force that acted as a dispersing agent, pushing the galaxies and stars away from each other in deep space; an energy that was repelling against the forces of gravity, forces that should pull everything together in what eventually would be the "Big Crunch"? Another recent observation also suggests the existence of Dark Matter, an invisible mass that would generate enough gravity to keep the rapidly spinning galaxies from flying apart. Existing "known" mass in galaxies such as our own Milky Way, does not have the amount of mass needed to keep stars on its outer edge, moving at great velocities, from flying off into space.

Review of Theory

Einstein and the other scientist at the turn of the 19th century were baffled by the fact that the speed of light was always the same velocity to an observer, no matter what the velocity of the observer or the velocity of the object emitting the light. Lorentz had shown that if an object was moving it contracted in the direction of its motion, increasing the distance light had to travel; this allowed the "perceived" speed of light to remain the same. Motion therefore warped space and time leading Einstein to develop his theories of Relativity.

The **Theory of Special Relativity** explains that motion is relative; Einstein showed this with the train and embankment thought experiment. Here two bolts of lightning strike simultaneously in front and behind a moving train. There is one observer midpoint between the lightning strikes on the moving train and one observer midpoint the lightning strikes on the embankment beside the train. Although the lightning strikes occurred at exactly the same time to the observer on the ground, the observer on the moving train saw the light from the front of the train first because of his forward motion. By the time the light meets this observer, he has moved towards the lighting in front of the train and away from the lightning behind the train.

In this text, we have reviewed a similar thought experiment with a train blowing its whistles simultaneously from the engine and caboose. We saw the different times the whistle was heard by the two observers and also saw the Doppler Effect by the movement of the train. Here, movement or motion caused a different reality for both observers.

The **Theory of General Relativity** went on to show that mass also has an effect on time and space. Mass causes space to bend around it, which in turn causes mass to attract mass, the effects of gravity. This also is what causes light waves to bend and slow down, manifesting itself in gravitational lensing and Red Shift. The effects of mass also cause clocks to change their tempo, in effect changing time. Therefore both mass and velocity have a direct influence on time and space. The greater the mass, the more space and time are warped and the greater the velocity (through acceleration in attaining that velocity), the greater time and space are warped.

Einstein in this theory, showed that gravity and acceleration are equivalent with his Equivalence Principle. This was shown with a thought experiment where an observer was floating in space in a large room with no gravity. A force was applied to the room that caused it to accelerate at a constant rate, forcing the observer onto the floor. The observer, not being able to see outside, assumed that this force was gravity. This asserted that the properties of gravity and acceleration are essentially the same thing.

Another thought experiment using multiple rocket stages in space can be done by comparing the velocity of one rocket stage to another; this experiment shows that velocity is relative to a reference point – one stage to another stage (Dennis, pg. 73).

This thinking was also shown with the baseball being thrown from a moving vehicle earlier in this text. The point here given was to show that if speed was relative, the effects of speed would also have to be relative. This means that since velocity changes time that time change would be relative to different observers at different relative speeds.

As mass increases at speeds close to the speed of light, this mass change must also be relative to different observers. An object would therefore be getting larger to one observer when compared to another observer at a different relative velocity; how can this be? Just as in Einstein's train, this can only be a **perceived** difference and not reality.

The only way that we can get to a point where the effects of velocity on mass becomes real, is to create a stationary point in space that all velocities are referenced to. The theory postulated here is that space is not empty and that a medium or field must be present. This medium not only propagates energy through "empty space" but also is the substance that is "curved" in empty space with various changes in densities and temperatures, creating the effect of gravity.

Gravity of the Matter

Let's Talk About Relativity

Let's look at this again. We all know the scenario where two cars are traveling toward each other; each car is traveling at 50 mph. That is 50 mph relative to the ground. Relative to each other they are traveling at 100 mph. As they pass, one passenger will throw a baseball to a passenger in the other car at 20 mph, relative to the car it is thrown from. Relative to the passenger catching the ball, say he had a radar detector aimed at the ball, the ball will be traveling at 120 mph. If an observer standing at the side of the road aimed a radar detector at the flying ball, it would read 70 mph (50 + 20). Of course here on earth we "usually" use the earth as the reference for speed. An aircraft on the other hand must use airspeed because the lift of the aircraft depends on it. In space however, outside our solar system, what is our reference for velocity?

In a thought experiment, let's put a satellite up in space and launch a rocket from it at 80,000 mps (miles per second). Let us now launch a smaller rocket from this rocket at 80,000 mps. Then another rocket from this rocket is launched at 80,000 mps. The third rocket should be traveling at 240,000 miles per second in reference to satellite from which the original rocket was launched.

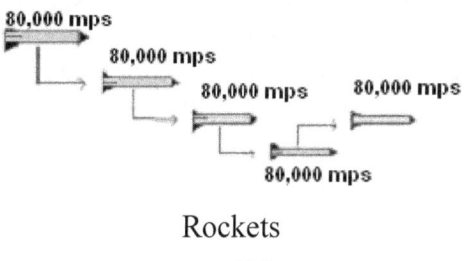

Rockets

Now what is the speed of light? 186,000 mps! What is it relative to? Electromagnetic energy (light, radio waves, x-rays) is always at the speed of light when observed, no matter what the position or velocity of the observer or receiver. There is no other reference point. The movement of any object is always measured by its change of position over a given period of time in reference to something; a change that may be in reference to its original position in space. Any object in space can be considered to be stationary unless it is changing in acceleration (speeding up or slowing down in reference to another point in space). In the previous illustration, all rocket stages are moving slower than the speed of light in reference to each other except, the first rocket launch sight (the satellite) relative to the last rocket. Neither the satellite nor the last rocket are accelerating or decelerating. The first may consider itself stationary or the last may consider itself stationary. The first rocket may be moving away from the last at a speed above the speed of light or the last rocket may be moving away from the first at a speed above the speed of light, which is it?

Einstein stated that an object's mass increases with its velocity. But what is its velocity relative to? He must be referring to stationary space! What is stationary space? How fast is our system moving relative to stationary space? If an object is moving at a high-speed relative stationary space, its mass increases. Does that mean that if an object slows down to zero velocity relative to stationary space, it will be non-existent and have no mass?

If there is no state as stationary space and the last rocket moves away from the first at or above the speed of light, then the last rockets mass must increase. If the last rockets mass increases, the first rockets mass becomes smaller relative to the last. Or is the first rocket moving away from last rocket at the speed of light? If this is the case, then the first rocket should be getting larger, not smaller, in relation to the last rocket.

I'm afraid reality is not as complex as it may seem. Let's be careful not to make something that we perceive to exist, to be absolute. Consider a clock on a rocket (Dennis, pg. 99). As the rocket moving away from us sends a radio signal of its clock, we perceive the clock to be ticking slower. As it returns, it is perceived to be faster. But reality is different. The clock never changed its rate of ticking. It never changed time!

This *medium*, the "aether", must exist. The whole electro-magnetic spectrum and the theory of gravity are dependent on it. Theoretically, this *medium* is mass and relatively stationary in undisturbed space. The smallest of all particles, it may consist of all three energy states, positive, negative and neutral. The neutral state may simply be a combination of both positive and negative states. The *medium* particles maintain an equidistant spacing from each other, depending on the density of the surrounding space. The *medium* is contained in both mass and empty space. Within mass and close to strong gravitational fields (caused by variances in density in this medium), the *medium* is much denser affects the transmission of electromagnetic waves through it.

195

Gravity of the Matter

The velocity of electromagnetic propagation depends on the density of the *medium*. In materials that have high density, the *medium* is denser, such as in glass and plastics, and the velocity of the energy through them is slightly slower. The *medium* is believed to be an elastic solid, unlike the assumptions of Michelson and Morley, and is compressed by moving objects, gravitational, electric and magnetic fields. It is in fact the change in density and pressure of this medium that generates these fields.

Outside of any movement and external fields, such as magnetic or gravitational fields, the *medium* falls to a stationary and neutral "pressure" dependent on its temperature. In front of a moving object, the *medium* compresses and therefore increases in "pressure". In the wake, it falls to negative pressure and expands; the "pressure", being the ratio of the spacing of its particles in reference to stationary, neutral space. All electromagnetic energy is transferred through the *medium* from an emitter to a receiver. The number of particles in a given space determines the velocity of the electromagnetic waves propagation; the denser the *medium*, the less elastic the *medium* becomes and slower the electromagnetic propagation.

Stephen Hawking, in his book, *A Brief History of Time*, suggests that there may be a form of matter that is uniformly distributed, yet still undetected that could raise the average density of the universe (Hawking, pg. 48). Could this be the medium that bends light from stars and slows down radar signals bounced off planets? Could this be the dust that causes Mercury to precess more than expected in its orbit close to the sun? Is this the force that changes the time in the GPS satellites or the extra mass contained in the galaxies that prevents the outer stars from flying out into the open universe? Is the variation in its density that causes Space-Time to warp and create the geometric effect we call gravity?

Gravity of the Matter

Conclusion

The **first hypothesis** we will make is that energy is the movement of mass. Scientifically speaking, energy is a variable quantity that is a characteristic of an object (mass). There are several ways to describe energy; kinetic, potential, gravitational, electromagnetic, chemical and nuclear. Energy can be transformed from one form of energy to another such as light into heat, but cannot be created nor destroyed. Action at a distance, such as the transfer of energy as light or gravity must therefore be transferred through the movement of mass or changes in its density; this may involve the photon and the graviton as taught in modern physics.

Light moves in waves and clearly generates the Doppler Effect when objects that give off energy resonate. Waves move through some sort of medium that is relatively stationary in space; an example is waves in water and sound through air or any other object such as steel. Here the **second hypothesis** is made that light waves must move through <u>something</u> that is relatively stationary in space; a point of reference for all matter.

Light waves are transmitted in both longitudinal and transverse waves. Longitudinal waves can move through any solid or fluid, however, transverse waves can only move through solids. The **third hypothesis** therefore states that the substance that light waves moves through must be an elastic solid – generating the problem of how planets and stars can move in rigid space.

If space is rigid, then all movement must consist of energy waves passing from one particle of this semi rigid substance to another. Standing waves manifest themselves normally as interference patterns generated in two or more waves of similar harmonics, interacting with each other in a stationary medium. Our **fourth hypothesis** is that matter may be simply manifested as energy waves moving through a solid medium. Clouds are an example of this; a lenticular cloud forming over a mountain ridge may seem to be created of stationary particles. This cloud however is formed by moving moist air flowing over a mountain top, condensing its moisture in the cold winds at high altitude as the air drops below the dew point. As the air moves down the downwind slope it warms up and evaporates. This constant movement of air creates a "standing wave", a cloud that seems to stand still, however generated by fast moving condensing then evaporating water particles.

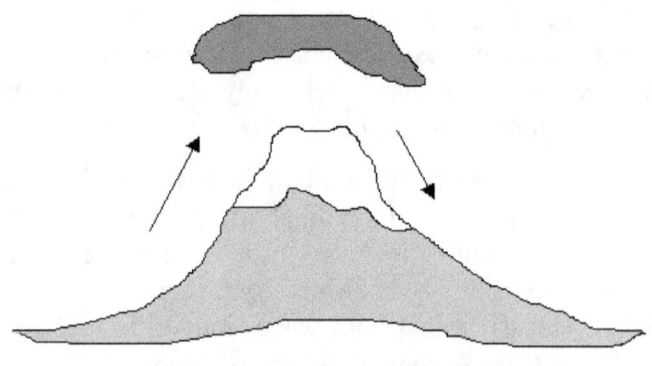

Lenticular Cloud

Conclusion

Another form of standing wave can be seen as vapor trails over the wing of an airplane, especially on a moist or rainy day. Here, similar to the lenticular cloud, moving air condenses in the low pressure region above the wing. Although this condensation seems to be stationary to the passengers in the aircraft, the air moving through this area may be moving at speeds over 500 miles per hour. Moisture in the moving air condenses momentarily in the low pressure area and then disappears as it moves away from the wing creating a standing wave.

Low Pressure Condensation

Similarly, photons and electrons may simply be the manifestations of confined energy. Energy moving as waves at the quantum levels must be transferred in packets of Plank's Constant and cannot be divided. Energies at these quanta will then move in a straight line or possibly in a confined, circular orbit – as if trapped; it cannot be dispersed. This trapped or confined energy then manifests itself as a moving particle in a "stationary" medium.

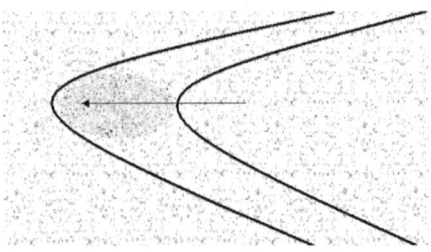

Photon Moving as a Particle in a Solid Medium

Energy of a photon may move from particle to particle in a stationary medium, yet confined to a specific size and area due to Planks Law. The particle moves in a straight line at the speed of light until absorbed by some other substance, normally an atom of some sort. A photon would therefore manifest itself as a moving particle even though it would simply be energy moving though a stationary solid substance as a standing wave; this would give it its wave properties.

An electron would likewise move through this medium as the photon does, however, it would only exist at specific regions known as shells around the atom. As it would possibly gain angular momentum, due to absorption of energy in the form of a photon, it will disappear and then reappear in another shell further out. Energy absorbed above a specific level will eject this extra energy as a photon away from the atoms nucleus. An electron would therefore manifest itself as a moving particle even though it would simply be confined energy moving though a stationary solid medium around a condensed particle.

Conclusion

It is possible however, that Keplerian dynamics and angular momentum do not pay a role as we may think with these particles. Since these particles may simply be the manifestation of energy in a "stationary" medium, there is no true net movement of mass, just waves of energy moving through a semi solid substance. The velocity of the electron wave may therefore increase in speed the further away it moves from the center of mass or atom and the medium becomes less dense. Toward the core, the closest electrons to the atom may move the slowest in the colder and denser medium contrary to Keplerian dynamics.

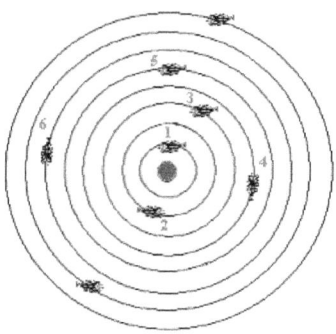

Electrons as Waves Moving Around a Nucleus

Without this stationary substance that light waves would move through, there would be no reference for the speed of light. Photons as moving particles would move relative to the object ejecting them or absorbing them, and light could not move at a constant velocity. Dual stars orbiting each other and moving in opposite directions as seen by Willem de Sitter, would give off light at different speeds to different observers. The mere fact that light generates Doppler shift is proof that light moves as a wave through a medium. Light or any other form of electromagnetic energy would therefore always move at a specific speed relative to the observer, similar to the speed of sound moving in the submarines sonar detectors dome or the speed of sound in the ear canal of someone hearing a trains whistle.

The **fifth hypothesis** will then state that other forms of mass would form from the cooling and coalescing of particles of this substance that is dispersed throughout the universe. This substance would contain hot and cold areas; cold areas having more densely spaced particles in this substance with a lower "pressure" due to low temperature. This pressure differential of low pressure close to particles and high pressure away from particles will cause particles to move together and the generation of gravity; the more the mass, the higher the differential pressures pushing mass together.

We can visualize this process by a simple thought experiment. Imagine if you will, a room separated in the middle by a thin plastic membrane, sealed off so that air cannot pass from one side of the room to the other. The room will be quite warm, say 38^0C or 100^0F and very humid. We then, using a heat exchanger with a refrigerant of some sort consisting of a condenser and evaporator, transfer heat energy from one side of the room to the other with no net air movement; both sides air tight and insulated. Over time, one side of the room will become cooler and cooler as the other side of the room becomes hotter and hotter. Soon the cold side of the room begins to fog up as water vapor starts to condense in the air and droplets of water begin to form on cooler surfaces; it doesn't take long until frost begins to form on the coldest objects in the room. Now, take notice of the thin plastic membrane, how it now bulges into the cool area of the room. Here, the hot air, having higher pressure, pushes in on the cold area, ballooning if you will, the membrane into the cold room; the hot region, pushing in on the cold region.

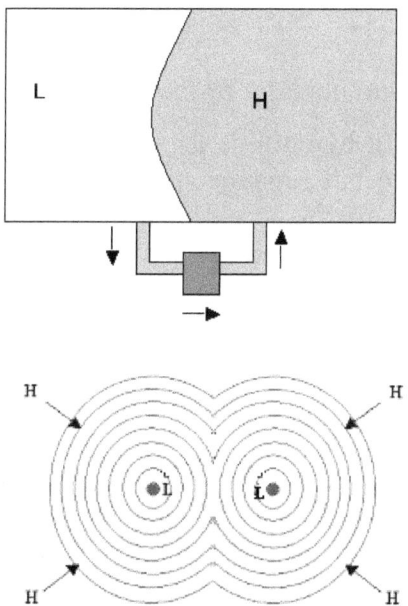

Back to the real world, energy such as light waves moving through these denser regions, would move slower and refract, bending the light, similar to how light moving from air into water in a glass, makes a spoon seem to bend. This creates a **sixth hypothesis** that the speed of light varies according to the density of the medium it passes through; moving slower in higher densities close to mass and faster in less dense areas far away from any mass.

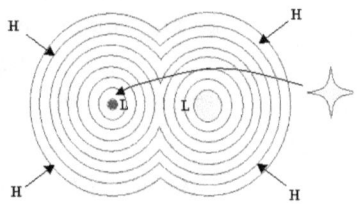

Starlight Bent by the Suns Gravity

The **seventh hypothesis** then states that if light moves slower close to large masses, that time its self will move slower. The opposite would then be true that the lack of gravity would in fact speed up time. Black holes, having the most mass therefore would have the largest effect on time to the point at the center where there is little or no energy, a singularity exists where time stands still. Time may vary from region to region, each area of space having its own "local time"; time however, could never be reversed, only stopped or sped up as time in affect is created by the movement of mass. Could these singularities be the portals to other dimensions?

Just as with sound in air or water, energy moving through this medium in space will move at a constant velocity depending on the "density" and "temperature" of the medium. Although nothing that consists of energy waves could move faster than the speed of light, the speed of light varies from one region of the universe to another depending on the density of this medium. This then would suggest the **eighth hypothesis** that wormholes do exist. Wormholes would be areas in the universe almost void of this medium. Energy moving in this area could move at velocities thousands of times faster than the speed of light in "near space" or space close to massive objects such as our sun. If mass merely consists of energy waves moving through this medium, then mass could very well be able to move through these areas at very high speeds, allowing one to "pop out" on the other side of the universe. A wormhole in effect would be a gravity tube in which an object would "fall" through due to an extreme imbalance in energy. Theoretically local wormholes could be used to "beam" someone to another location relatively nearby; galactic travel utilizing such a method could also be developed and light speed travel or "warp speed" not necessary.

If this medium is dispersed throughout the universe in all directions and was created by the "Big Bang", it then should become less dense, the further away from the center of the source of the energy, to the point where it no longer exists. At this point energy can no longer propagate (similar to how sound waves cannot propagate out into space) and matter and time can no longer exist. The existence of the edge of the universe is therefore the **ninth hypothesis**.

The **tenth hypothesis** is that this universe is slowly collapsing. Edwin Hubble's discovery in 1929 suggested that galactic Red Shift in light from stars and galaxies increases in proportion to their distance from earth. His studies showed that 38 of 40 galaxies observed are rapidly moving away from each other, an omni-directional galactic expansion. Galaxies furthest away are moving the fastest – some at 57,000 kilometers per second or 13 million miles per hour – 2% the speed of light. The problem here is that this is evidently omni-directional – indicating that we are the center of the universe. If we were viewing this expansion out on the edge, only certain areas focused in one region would have this Red Shift; either we are the "center of the universe" or this observed redshift, therefore, is simply an illusion; an illusion due to the shrinking of the universe around us as it cools, loosing energy out into the vastness of the universe.

Since it takes time for the universe to cool and also takes time for light to reach us on earth, the changes to the universe will be seen uniformly in all directions and be relatively consistent at various distances. This however suggests a possible "Big Crunch" as energies subside and particles coalesce. Eventually as the medium between all objects condenses and cools, there will be nothing to keep objects apart, this expansion medium will no longer exist. Higher energies outside will eventually push all matter together into one place. Here atoms can no longer exist, fusing together into what has been called the **Bose–Einstein Condensate** or BEC's, huge molecules that form at near 0 Kelvin or absolute zero and possibly what Black Holes are made of.

Summary

In summary, gravity has been shown to be a geometric effect of curved Space - Time and is not a propagated source of energy as we see in electromagnetic and sound waves. The effects of gravity are therefore "local" and instantaneous. This has been shown with the simple principle of aberration of sunlight, how that the perceived position of the sun in the sky is approximately 8.3 minutes behind its true position due to the transit delay in light waves through space. Gravitation has no such aberration or delay; research has shown that the gravitational position of the sun in reference to the earth in its orbit around the sun is the same as it true position. If this were not the case, the earth would orbit around the aberration and not the true position of the sun, eventually resulting in the decay of the orbit. If gravity was propagated in waves and traveled the great distances of space over time, orbits of bodies could not be maintained and the universe would break apart.

Imagine in a thought experiment a steel girder being hoisted by a crane into position on a skyscraper with a line of dominoes on top of the girder. As one end of the steel girder impacts its positioning point, the dominoes begin to fall. Domino's fall at approximately 1 meter per second therefore taking about 50 seconds to travel the length of a 50 meter girder. Although the speed of the dominos takes only 50 seconds to travel the length of the beam, the beam itself stops its forward motion immediately. In much the same way, gravity like the girders motion occurs immediately whereas the light waves like the dominoes takes time to propagate over a distance.

Another analogy that can be used it that of helium filled balloons. A balloon begins to rise when it is released because the balloon is "lighter" than air. The Ideal Gas Law which was developed by Emile Clapeyron in the early 1800's. This law was established in order to explain the properties of gasses in relation to each other and is expressed as $pV = (w/mw) \times (RT)$ or simplified as $pV = nRT$. The p in the equation represents "pressure," V represents "volume" in "moles," w/mw represents weight (gravitational force) divided by molecular weight (the substance of the gas expressed as n), R is the gas constant and T represents the temperature of the gas.

This formula works for buoyancy of objects in all fluids such as a piece of wood or lead pipe in water. The Ideal Gas Law states that an object will "float" in a fluid due to the fact that the force exerted on the object by the fluid around it is greater than the force of gravity on that object. This formula takes gravity into account in order to calculate the buoyancy of an object and cannot be used to calculate gravity itself. However the same principle can be used; an object moves in a specific direction because it is less dense or denser than the fluid surrounding it. Therefore, if an object in a fluid has a fluid denser in one direction and less dense in the other, it will move toward the area which is less dense.

This very same principle can be seen in the weather formation of storm systems on earth, especially in the spin and force of a hurricane. The core of a low pressure area consists of air that is less dense that the air in a high pressure area. This creates a pressure differential that causes air to flow as "wind" in toward the low pressure areas; the differences in pressure ranges from low to high as pressures are measured out from the center of the system which can be charted as pressure gradients or isobaric lines. A sailboat on a lake will be blown and move toward the center of the low pressure system as the air flows from a denser region to a less dense region. The Gravity of the Matter can be better seen through this hypothesis.

Our **first** hypothesis is that energy cannot exist without mass, much like velocity cannot exist without the bullet or speed cannot exist without the car. Pure energy by itself cannot exist; mass cannot become energy and energy cannot become mass. Energy is simply manifested as the movement of mass in some way or form; energy is a property of mass.

Second, we hypothesize that light waves must move through some sort of medium, much like waves move through water or air. In order to have a wave, some sort of medium or field must be waving; something like what was once called the aether. The study of the movement of galaxies suggests that something else exists within these galaxies that tie all the stars in them together, causing them to move in unison counter to Keplerian dynamics. It is now believed that Dark Matter is this component that ties the stars together within the galaxies. Dark Energy is also now theorized in order to support Hubble's Law and the apparent expansion of our universe and the force that pushes galaxies apart. Today it is believed that the universe is composed of 73% of Dark Energy, 23% of Dark

Matter, and 4% ordinary Baryonic matter of which we are made of. Could this Dark Matter and Dark Energy be the manifestation of this medium?

On the 4th July 2012, the Higg's Boson, also known as the "God particle" was discovered at the Large Hadron Collider at CERN (Conseil Européen pour la Recherche Nucléaire) near Geneva Switzerland. Although much more complicated then said here, this particle was only theorized to exist for some forty years. It is a particle that has no spin, no electric charge and exists in "empty space" as the Higgs Field and is believed to be the mechanism in which particles acquire mass. It was the missing link in the physics Standard Model theory which was needed to explain how the interactions of the electromagnetic, weak, and strong nuclear forces work. Could this be the very medium in which matter is hypothesized here to be manifested in through the energies of Standing Waves?

The **third** hypothesis is that whatever light or electromagnetic energy is moving through (the medium or aether as it might be called) must be considered to be an elastic-solid. This is due to the fact that transverse waves, waves that are polarized, can only move through solids and cannot move through fluids. Light is an electromagnetic wave that clearly is polarized and transmitted as transverse waves.

The **fourth** hypothesis is that mass may consist of standing waves within this elastic-solid medium. This is why the planets, stars and galaxies are not rigidly in place and move freely within this medium. Picture the universe,

Summary

if you will, as a three-dimensional television screen, where individual pixels close up, do not move. Simply speaking, the energy that is changing the pixels color and brightness or amplitude, moves from pixel to pixel, giving the illusion of movement while the pixels remain solidly in place. These pixels simply seem to appear and disappear, jumping from one place to another without net movement.

The **fifth** hypothesis is that particles of the subatomic world are this illusion; an illusion created by energy moving through this medium which is created by the cooling and coalescing of the medium and manifested by confined pressure waves moving through it. This is why the Michelson Morley experiment failed; mass is the medium, manifested by energy moving through it. The pressure differentials in this medium therefore create the illusive condition called gravity. Here, higher energies that are remnants of the Bing Bang, far away from the coalescing particles of mass, press in on the lower energies of the colder condensed mass (including the fiery stars), causing mass to attract mass. Due to these pressure differentials existing simultaneously as pressure gradients around mass, gravity occurs instantaneously and has no delay.

Our **sixth** hypothesis states that closer the medium is to the coalescing mass, the denser the medium becomes, causing light passing through this region of space to change in velocity. Since light moves in waves, refraction occurs as it passes through these denser plains and causing light waves to bend as they pass close to large masses. This was evident in Eddington's discovery of starlight bending as it passed close to the sun's surface.

213

Our **seventh** hypothesis is that time will vary with the strength of gravitational fields, creating what is called "local time." Time is perceived by the changes in the position mass due to energy moving through the medium, frame by frame; if not change occurs in these frames, time is standing still. Energy moves slower closer to mass due to the mediums increased density around it. Time therefore will move slower the closer an object sensitive to time such as a clock comes to mass. Time also will vary during acceleration and deceleration due to the "principle of equivalence" where gravity and acceleration are indistinguishable; it is not therefore velocity that changes time, it is the acceleration or deceleration to that velocity of an object that changes "local time" for that object.

Our **eighth** hypothesis is that wormholes do exist. In the far reaches of the universe where the mediums density is almost void, energy may move at velocities much higher than our calculated speed of light. Since mass is the movement of energy through this medium, mass in this region may simply exist, then not exist, and then exist again, "jumping" as it were, far from where it came from. This jump is almost identical to the jump the electron encounters as it jumps from one shell to another, existing, then not existing, then existing again.

This brings us to our **ninth** hypothesis that at some point far away from the center of the "Big Bang" a place exists where space is completely void of this medium that mass can no longer exist. If mass is manifested as standing waves through this medium, mass and therefore also time cannot exist outside of this medium, just as sound waves cannot exist in space. This may not only occur just at the edge of Time-Space but may also come into play within the vast areas between the galaxies in currents of nothingness. Mass as it would enter these areas would simply "jump" across this emptiness as if entering a wormhole.

Small wormholes may also be created here at home, allowing one to "beam" from one place to another. This could be done by "energizing" or heating this medium in close vicinity to a transport vehicle much like how we heat gasses to accelerate rockets today. Energizing the medium would create pressure differentials in the medium similar to gravitational forces explained earlier. Since the medium that is being energized or heated, consists of so little mass, it may possibly take very little energy in order to create this force; much like the energy causing a balloon to float.

It may simply be a matter of frequency, similar to a microwave oven heating your diner. Using some sort of heat exchanger where cooling the medium in front of a transport vehicle and taking that energy to heat the medium behind the vehicle could induce this transport. Another possible form of propulsion could be inducing an imbalance in pressure magnetically with some sort of flux flow.

This technology may allow a craft to hover or move at extreme velocities without the negative effects of G forces on its occupants, as if "falling" toward its intended destination and possibly "falling up" into the sky.

Our **tenth** hypothesis is that the universe is slowly collapsing. Red Shift seen from galaxies furthest away from us suggests that rather than having an expanding universe, the universe around us is slowly collapsing. As energy in our universes core is radiated out into the outer reaches of space, energy is being lost and our core is cooling and shrinking.

As we look out into space we are looking back into time. As we look back into time, an illusion possibly is being created that suggests that these galaxies are racing away from us omni directionally. The fact is that it could be that we are racing away from these galaxies, as if falling into a "black hole" as the core of the universe around us quickly cools. Black holes are all around us and are probably at the centers of all galaxies and even in regions of dark "empty" space. Could it be that black holes are regions where the universe is gone completely cold?

Summary

Where energy in mass has been completely "wrung out" and the "Bose Einstein Condensate" is created? Is this the beginning of the "Big Crunch"?

It is in understanding the Gravity of the Matter that surrounds us, that we may possibly someday soon lose the hold that gravity has on us; being able to move across space untethered and unchained by its grasp. Transport to far distant stars and planets in other galaxies may soon be realized. We are living in the early twenty-first century, almost 100 years after Einstein's theories were developed. These theories became the foundation of quantum mechanics and modern theoretical physics that is taught in our schools today. Is it not time to begin questioning these old theories again?

"To deny the aether is ultimately to assume that empty space has no physical qualities whatever. The fundamental facts of mechanics do not harmonize with this view..."

Albert Einstein (Isaacson, pg. 318)

Gravity of the Matter

References

Allen, Tim (1996) I'm Not Really Here, Hyperion, 114 Fifth Avenue, New York, NY

Arago, François (1858) Oeuvres Complètes, 7th Edition, Volume 4, Paris

Aristotle (350 B.C.E) De Caelo, Book I, Chapter 2

Ashtekar, Abhay (2005) One Hundred Years of Relativity, World Scientific Publishing, 5 Toh Tuck LinkSingapore

Asimov, Isaac (1991) Atom, Journey Across the Subatomic Cosmos, Truman Talley Books, First Plume Printing, New York, NY

Barbour, Julian B. (2001) End of Time: The Next Revolution in Physics, Oxford University Press, 198 Madison Avenue New York, NY

Bartusiak, Marcia (2000) Einstein's Unfinished Symphony, Listening to the Sounds of Spac^{2}e -Time, Burkley Publishing Group, New York, NY

Bless, Robert C. (1996) Discovering the Cosmos, University Science Books, 55D Five Gates Rd, Sausalito CA

Bord, Donald J. (2000) Inquiry into Physics, Brooks / Cole Fourth Edition.

Borel, Armand (2001) Oeuvres, Collected Papers, Vol. IV, Springer Scientific and Business Media, 1150 65th Street, Emeryville, CA

Born, Max (1962) Einstein's Theory of Relativity, Dover Publications Inc. New York, NY

Brown, Julian (2000) Minds, Machines, and the Multiverse, The Quest for the Quantum Computer, Simon and Schuster, New York, N.Y.

Collins, Harry (1998) The Golem: What You Should Know about Science, Cambridge University Press, Cambridge CB2 2RU, UK

Dekel, Avishai (1999) Formation of Structure in the Universe, Cambridge University Press, Cambridge CB2 2RU, UK

Dennis, Mark (2010) A Challenge to Einstein, Createspace Self-Publishing, Createspace.com

Ditchburn, R.W. (1991) Light, Dover Publications, Inc., 31 E.2nd Street, Mineola, NY

Einstein, Albert, (1952) Relativity by Einstein, Special and General Theory, Tess Press, Fifth Edition

Evans, Alvis, J. (1985) Basic Electronics Technology, Tandy Corporation, Ft. Worth, Texas

Fresnel, A. (1818) Lettre d'Augustin Fresnel à François Arago Annales de chimie et de physique. 9, pg. 57–66

Galileo, Galilei (1632) Dialogue Concerning the Two Chief World Systems, Second Edition, University of California Press, Berkeley, CA

References

Glendenning, Norman K. (2007) Our Place in the Universe, World Scientific Publishing, 27 Warren Street, Hackensack, NJ

Hawking, Stephen (1988) A Brief History of Time, Bantam Books, Random House Inc. New York, N.Y.

Hawking, Stephen (1993) Hawking on the Big Bang and Black Holes, World Scientific Publishing, pg. O. Box 128, Farrer Rd., Singapore

Hobson, Michael P. (2006) General relativity: An Introduction for Physicists, Cambridge University Press, Cambridge University Press, Cambridge CB2 2RU, UK

Isaacson, Walter (2007) Einstein, His Life and Universe, Simon and Schuster Paperbacks, New York, NY

Kafatos, Menas (1990) The Conscious Universe, Springer-Verlag, New York, NY

Lorentz, H.A, (1909) The Theory of Electrons, B.G. Teubner, G. E. Stechert, Booksellers, New York, 129-133 West 20th St. New York, NY

Newton, Sir Isaac, (1730) Opticks, a Treatise of the Reflections, Refractions, Inflections and Colors of Light, Fourth Edition.

Pitaeskii, Lev (2003) Bose-Einstein Condensation, Oxford University Press, USA, 198 Madison Avenue New York, NY

Poincaré, H. (1904) The Present and the Future of Mathematical Physics, Address Delivered to the International Congress of Arts and Science, St Louis.

Polhinghorne, J.C. (1984) The Quantum World, Princeton Science Library, Princeton University Press, Princeton NJ

Psillos, Stathis (1999) Scientific Realism, Routledge, 29 W. 35th Street, New York, NY

Ross, W.D. (1927) The Works of Aristotle: Translated into English Under the Editorship of W.D. Ross, M.A., the Oxford University Press, London UK

Sokolov G., (1999) The Theory of Relativity and Physical Reality

Tippler, Paul (2003) Physics for Scientists and Engineers, Standard Version, W. H. Freeman Company, 41 Madison Avenue, New York, NY

Van der Kamp, Walter (1988) De Labore Solis, Airy's Failure Reconsidered, Anchor Book & Printing Centre, Surrey, B.C., Canada

Veatch, Henry(1978) Electric Circuit Action, Science Research Associates, Chicago ILL

Wolf, Fred, (1989), Taking the Quantum Leap, The New Physics for the Non-Scientists, Harper and Row Publishers, New York, NY

Appendix

I cannot but regard the aether, which can be the seat of an electromagnetic field with its energy and its vibrations, as endowed with a certain degree of substantiality, however different it may be from all ordinary matter (Borel, pg. 690).

(Lorentz, 1906)

Indeed one of the most important of our fundamental aether not only occupies all space between molecules, atoms, or electrons, but that it pervades all these particles. We shall add the hypothesis that, though the particles may move, the aether always remains at rest (Lorentz, pg. 11).

(Lorentz, 1906)

Lorentz proclaimed the very radical thesis, which had never before been asserted with such definiteness: the aether at rest in absolute space.

In principle this identifies the aether with absolute space. Absolute space is no vacuum, but something with definite properties whose state is described with the help of two directed quantities, the electrical field E and the magnetic field H, and as such, it is called the ether (Born, pg. 204).

(Born, 1924)

Gravity of the Matter

Index

Index

www.ingramcontent.com/pod-product-compliance
Lightning Source LLC
Chambersburg PA
CBHW051449170526
45166CB00001B/175